TARGETING U.S. TECHNOLOGIES:

A Trend Analysis of Cleared Industry Reporting

2014

DSS Mission
DSS supports national security and the warfighter, secures the nation's technological base, and oversees the protection of sensitive and classified information and technology in the hands of industry.
We accomplish this mission by: clearing industrial facilities, personnel, and associated information systems; collecting, analyzing, and providing threat information to industry and government partners; managing foreign ownership control and influence in cleared industry; providing advice and oversight to industry; delivering security education and training; and, providing information technology services that supports the industrial security mission of the Department of Defense and its partner agencies.

Date of Information: October 1, 2013

Product Coordinated with: ACIC, AFOSI, and NGA

PREFACE

The 11th-century Danish ruler King Canute was the commander of forces that had made him ruler of England, Denmark, Sweden, and parts of Norway. The king found himself surrounded by sycophants who continually told him how powerful he was. So he ordered his throne carried to the seashore, whereupon he sat down on it. Then he commanded the waves to cease their advance. As the tide continued to rise, Canute's feet, shoes, and chair became wet. He then turned to his courtiers and reproved them for their unjustified flattery. Not even a royal command could halt the tides.

We might wish that we could simply brandish a scepter or issue a command and bring to a halt the wave of attacks on our "shores"—the U.S. cleared industrial base. Seemingly like an inexorable tide, reported foreign collection attempts to obtain unauthorized access to sensitive or classified information and technology resident in the U.S. cleared industrial base continue to rise. In fiscal year 2013, there were over 30,000 reports of suspicious activity, and the number of suspicious contact reports industry submitted concerning foreign collection attempts increased by 33 percent from the previous year.

As King Canute knew, and as his seashore demonstration to his courtiers clearly testified, it takes more than wishing or simply issuing orders to stop ill tides from submerging us. Preventing illicit foreign collection attempts from blunting the edge of our military's sword or striking it from our grasp, thereby endangering our warriors as they go into battle, requires a combined effort between industry, the intelligence and law enforcement communities, and other government entities.

The Defense Security Service (DSS) oversees the protection of sensitive or classified information and technology resident in the U.S. cleared industrial base. This annual publication, *Targeting U.S. Technologies: A Trend Analysis of Cleared Industry Reporting*, represents part of the effort to maximize the effectiveness of those endeavoring to maintain our national security. It is cleared contractors who most directly feel the hostile waters rising around their feet; amidst their efforts to fend them off, they report to DSS. DSS builds on reports from industry to develop analytical assessments that articulate the threat to U.S. information and technology in cleared industry.

DSS collects these reports, evaluates and analyzes them regarding both continuities with and departures from previous patterns, and publishes the results. This process benefits our national security, warfighters, cleared industry partners, and local communities. The information in this report helps employees, companies, and intelligence and law enforcement professionals better understand the nature of the hostile influences that so persistently oppose us. Increased awareness of which U.S. technologies foreign entities are targeting and the methods of operation they use in their attempts to acquire them can only make us better at identifying and thwarting illicit collection activity. In fiscal year 2013, our combined efforts produced 717 operations or investigations based on information industry provided. Over 80 percent of these are still under way, with many foreign collectors already identified, isolated, diverted, or otherwise thwarted.

Foreign collectors are not water: elemental, eternal, and unconcerned with human endeavors. Rather, they are human beings whose immediate and ongoing efforts to obtain unauthorized access to U.S. information and technology demonstrate the ever-changing advances that flow from misguided human ingenuity. They are also, at the least, underhanded, and frequently hostile. Foreign collectors are not ocean waves, responding to relentless but natural forces of winds and tides. They are foreign governmental, quasi-governmental, or commercial agents or private individuals who make conscious decisions to illicitly seek a shortcut to out-fighting U.S. warfighters or out-competing U.S. companies.

Foreign collectors' efforts can be thwarted, but only through a team effort among cleared contractors, DSS, and intelligence and law enforcement partners. Constant and better-attuned vigilance, smarter methods and defenses, and increased and mutual support can help our cleared industrial base keep our country clear of the encroaching waters, and thus strong, healthy, and secure.

Stanley L. Sims
Director
Defense Security Service

CONTENTS

FIGURES

THE CHARGE TO THE DEFENSE SECURITY SERVICE

The Defense Security Service (DSS) supports national security and the warfighter, secures the nation's technological base, and oversees the protection of U.S. and foreign classified information in the hands of industry. The DSS Counterintelligence (CI) Directorate seeks to identify unlawful penetrators of cleared U.S. industry and stop foreign collection attempts to obtain unauthorized access to sensitive or classified information and technology resident in the U.S. cleared industrial base. DSS CI articulates the threat for industry and U.S. government leaders.

THE ROLE OF INDUSTRY

In carrying out its mission, DSS relies on the support of cleared contractor employees and the U.S. intelligence and law enforcement communities. Chapter 1, Section 3 of the *National Industrial Security Program Operating Manual (NISPOM)*, 5220.22-M, dated February 28, 2006, requires cleared contractors to remain vigilant (in person and online) and report any suspicious contacts to DSS. The process that begins with initial reporting of all such contacts and continues with ongoing and collective analysis reaches its ultimate stage in successful investigations or operations.

In accordance with the reporting requirements laid out in the *NISPOM*, DSS receives and analyzes suspicious contact reports from cleared contractors. DSS categorizes these reports as suspicious, unsubstantiated, or of no value. For each reported collection attempt, DSS data aggregation and analysis methodologies seek to gather as much information as possible: who instigated the attempt, where it came from, what its aim was, and what methods of collection it used. Comprehensive analysis of information from across all companies and elements of cleared industry forms the basis for this report, and determines the actions DSS takes and the advice it gives to cleared contractors to combat the threat.

Cleared contractor reporting provides information concerning actual, probable, or possible espionage, sabotage, terrorism, or subversion activities to the Federal Bureau of Investigation and DSS. When indicated, DSS refers cases of CI concern to its partners in the law enforcement and intelligence communities for potential exploitation or neutralization. DSS follows up by informing industry of remedial actions that can decrease the threat in the future. This builds awareness and understanding of the individual and collective threats and actions and informs our defenses.

Defense Security Service Reporting Definitions

The Defense Security Service (DSS) sorts each report received from cleared industry under the National Industrial Security Program Operating Manual Section 1-302b into one of three distinct categories: suspicious contact report (SCR), unsubstantiated contact report (UCR), or assessed no value (ANV). Subsequent information and reevaluation may cause changes in these categorizations, e.g., an SCR may change to a UCR.

Suspicious Contact Report (SCR) – A report DSS receives from cleared industry that contains indicators that it is almost certain or likely or there is an even chance that some individual, regardless of nationality, attempted to obtain unauthorized access to sensitive or classified information and technology or to compromise a cleared employee. Reports designated as SCRs represent incidents most likely to have involved actual attempts to do so.

Unsubstantiated Contact Report (UCR) – A report of an incident in which it is unlikely that any individual, regardless of nationality, attempted to obtain unauthorized access to sensitive or classified information and technology or compromise a cleared employee. However, DSS retains such reports, as the aggregate of several UCRs or information obtained subsequently may result in the identification of foreign intelligence activity.

Assessed No Value (ANV) – A report that only remotely represents a CI concern, such as an email or credit card scam. DSS does not retain reporting assessed as ANV.

THE REPORT

Department of Defense (DoD) Instruction 5200.39, Critical Program Information Protection within the DoD, dated July 16, 2008, requires DSS to publish a classified and an unclassified report, each detailing suspicious contacts occurring within the cleared contractor community. The focus of this report is on efforts to compromise or exploit cleared personnel, or to obtain unauthorized access to sensitive or classified information or technologies resident in the U.S. cleared industrial base.

Each year DSS publishes *Targeting U.S. Technologies: A Trend Analysis of Cleared Industry Reporting*. In this report, the 16th annual *Targeting U.S. Technologies* (or *Trends*), DSS provides a snapshot of its findings on foreign collection attempts. It provides a statistical and trend analysis that covers the most prolific foreign collectors targeting the cleared contractor

community during fiscal year 2013, compares that information to the previous year's report, and places that comparison into a larger context.

DoD Instruction 5200.39 requires DSS to provide its reports to the DoD CI community, national entities, and the cleared contractor community. This report constitutes part of DSS' ongoing effort to assist in better protecting the cleared industrial base by raising general threat awareness, encouraging the reporting of incidents as they occur, identifying specific technologies at risk, and applying appropriate countermeasures. DSS intends the report to be a ready reference tool for security professionals in their efforts to detect, deter, mitigate, or neutralize the effects of foreign targeting. DSS previously released a classified version of this report.

Figure 1: Method of Operation Definitions

Academic Solicitation
Via requests for, or arrangement of, peer or scientific board reviews of academic papers or presentations, or requests to study or consult with faculty members, or applications for admission into academic institutions, departments, majors, or programs, as faculty members, students, fellows, or employees

Suspicious Network Activity
Via cyber intrusion, viruses, malware, backdoor attacks, acquisition of user names and passwords, and similar targeting, these are attempts to carry out intrusions into cleared contractor networks and exfiltrate protected information

Attempted Acquisition of Technology
Via agency of front companies or third countries or direct purchase of firms, these are attempts to acquire protected information in the form of controlled technologies, whether the equipment itself or diagrams, schematics, plans, spec sheets, or the like

Seeking Employment
Via résumé submissions, applications, and references, these are attempts to introduce persons who, wittingly or unwittingly, would thereby gain access to protected information that could prove useful to agencies of a foreign government

Request for Information
Via phone, email, or webcard approaches, these are attempts to collect protected information under the guise of price quotes, marketing surveys, or other direct and indirect efforts

Solicitation or Marketing Services
Via sales, representation, or agency offers, or response to tenders for technical or business services, these are attempts by foreign entities to establish a connection with a cleared contractor vulnerable to the extraction of protected information

Foreign Visit
Via visits to cleared contractor facilities that are either pre-arranged by foreign contingents or unannounced, these are attempts to gain access to and collect protected information that goes beyond that permitted and intended for sharing

Exploitation of Relationships
Via established connections such as joint ventures, official agreements, foreign military sales, business arrangements, or cultural commonality, these are attempts to play upon existing legitimate or ostensibly innocuous relationships to gain unauthorized access

Surveillance
Via visual, aural, electronic, photographic, or other means, this comprises systematic observation of equipment, facilities, sites, or personnel

Criminal Activities
Via theft, these are attempts to acquire protected information with no pretense or plausibility of legitimate acquisition

Search/Seizure
Via physical searches of persons, environs, or property or otherwise tampering therewith, this involves temporarily taking from or permanently dispossessing someone of property or restricting his/her freedom of movement

Figure 2: Collector Affiliation Definitions

Commercial
Entities whose span of business includes the defense sector

Government Affiliated
Research institutes, laboratories, universities, or contractors funded by, representing, or otherwise operating in cooperation with a foreign government agency, whose shared purposes may include acquiring access to U.S. sensitive, classified, or export-controlled information

Individual
Persons who, for financial gain or ostensibly for academic or research purposes, seek to acquire access to U.S. sensitive, classified, or export-controlled information or technology, or the means of transferring it out of the country

Government
Ministries of Defense and branches of the military, as well as foreign military attachés, foreign liaison officers, and the like

Unknown
Instances in which no attribution of a contact to a specific end user could be directly made

SCOPE/METHODOLOGY

DSS considers all reports collected from the cleared contractor community. It then applies analytical processes to them, including the DSS foreign intelligence threat assessment methodology. After sorting all reports into the three categories—suspicious contact report (SCR), unsubstantiated contact report (UCR), and assessed no value—we base this publication on SCRs and select UCRs. The analyses also incorporate references to all-source Intelligence Community (IC) reporting.

The *Trends* is organized first by region, then by collector affiliation, methodologies employed, and technologies, including the specific targeted sectors. It incorporates statistical and trend analyses on each of these areas. Each section also contains a forecast of future activities against the cleared contractor community, based on analytical assessments.

To organize its targeting analysis, DSS applies a system of categories and subcategories that identify and describe technologies. In FY13, DSS ceased analyzing foreign interest in U.S. defense technology in terms of the 20 sectors of the Militarily Critical Technologies List (MCTL), shifting instead to the 29 sectors of the Industrial Base Technology List (IBTL). Like the MCTL, the IBTL is a compendium of the science and technology capabilities under development worldwide that have the potential to significantly enhance or degrade U.S. military

capabilities in the future. However, the newer system updates the categorization scheme to incorporate emergent and cutting-edge technologies. In addition, it breaks out into separate categories some technologies that formerly were lumped together in larger sectors. The most significant changes affected the MCTL's information systems and lasers, optics, and sensors sectors. See the full IBTL at the end of this report for a comparison of the MCTL and IBTL.

This publication also refers to the Department of Commerce's Entity List. This list provides public notice that certain exports, re-exports, and transfers (in-country) to entities included on the Entity List require a license from the Bureau of Industry and Security. An End-User Review Committee (ERC) annually examines and makes changes to the list, as required. The ERC includes representatives from the Departments of Commerce, Defense, Energy, State, and, when appropriate, Treasury.

ESTIMATIVE LANGUAGE AND ANALYTIC CONFIDENCE

DSS employs the IC estimative language standard. The words of estimative probability used, such as *we judge*, *we assess*, or *we estimate*, and terms such as *likely* or *indicate*, represent the agency's effort to convey a particular analytical assessment or judgment.

Because DSS bases these assessments on incomplete and at times fragmentary information, they do not constitute facts nor provide proof, nor do they represent empirically based certainty or knowledge. Some analytical judgments are based directly on collected information, others rest on previous judgments, and both types serve as building blocks. In either variety of judgment, the agency may not have evidence showing something to be a fact or that definitively links two items or issues.

Intelligence judgments pertaining to likelihood are intended to reflect the approximate level of probability of a development, event, or trend. Assigning precise numerical ratings to such judgments would imply more rigor than the agency intends. The chart below provides a depiction of the relationship of terms used to each other.

Remote	Very Unlikely	Unlikely	Even Chance	Probably, Likely	Very Likely	Almost Certainly

The report uses *probably* and *likely* to indicate that there is a greater than even chance of an event happening. However, even when the authors use terms such as *remote* and *unlikely* they do not intend to imply that an event will not happen. The report uses phrases such as *we cannot dismiss*, *we cannot rule out*, and *we cannot discount* to reflect that, while some events are unlikely or even remote, their consequences would be such that they warrant mentioning.

DSS uses words such as *may* and *suggest* to reflect situations in which DSS is unable to assess the likelihood of an event, generally because relevant information is sketchy, fragmented, or nonexistent.

In addition to using words within a judgment to convey degrees of likelihood, DSS also assigns analytic confidence levels based on the scope and quality of information supporting DSS judgments:

High Confidence

- Well-corroborated information from proven sources, minimal assumptions, and/or strong logical inferences
- Generally indicates that DSS based judgments on high-quality information, and/or that the nature of the issue made it possible to render a solid judgment

Moderate Confidence

- Partially corroborated information from good sources, several assumptions, and/or a mix of strong and weak inferences
- Generally means that the information is credibly sourced and plausible but not of sufficient quality or corroborated sufficiently to warrant a higher level of confidence

Low Confidence

- Uncorroborated information from good or marginal sources, many assumptions, and/or mostly weak inferences
- Generally means that the information's credibility or plausibility is questionable, or that the information is too fragmented or poorly corroborated to make solid analytic inferences, or that we have significant concerns or problems with the sources

EXECUTIVE SUMMARY

Fiscal year 2013 (FY13) saw a continuation of the past decade's steady rise in reported foreign collection attempts to obtain unauthorized access to sensitive or classified information and technology resident in the U.S. cleared industrial base. While industry reports to the Defense Security Service (DSS) in FY13 increased a considerable 33 percent, this was less than the 50, 74, and 60 percent increases of immediately preceding years.

The six collector regions remained in the same relation to each other in FY13 as in FY12 with regard to the frequency with which they appeared in industry reporting to DSS. Increases for the six regions ranged from 15 percent for the relatively large number of East Asia and the Pacific reports to over 100 percent for the relatively small number of Western Hemisphere reports. With regard to shares of total reporting, the perennial top collector region, East Asia and the Pacific, receded six percentage points, while each of the other major regions gained between two and four percentage points.

While the number of industry reports linked to East Asia and the Pacific increased from FY12, they did so by only 15 percent, less than in recent years, and regional collector entities' share of the total decreased from 50 to 44 percent. A plausible explanation is that beginning in February 2013 press reports highlighted corporate and industry reports detailing a multi-year analysis of cyber espionage conducted by East Asia and the Pacific cyber operatives. The coverage included specifics concerning their infrastructure, protocols, and methods, and DSS' U.S. government partners followed up by providing industry with additional technical indicators. The level of industry reporting of cyber operations linked to East Asia and the Pacific dropped. But after a period of several months—sufficient for cyber actors to make adjustments—the level of reporting began to rise again.

However, the temporary suppression had an impact on other aspects of FY13 data. Suspicious network activity (SNA), the method of operation (MO) most commonly reported in recent years for collection attempts linked to East Asia and the Pacific, decreased in number of cases and slid from 29 to 19 percent of the yearly total. This publication discusses these issues in greater detail in the East Asia and the Pacific section.

The temporary overall decrease in reports of SNA allowed academic solicitation, which increased over 80 percent in number of reported cases year-over-year, to become the most common MO in overall reporting at 22 percent. A similar MO, seeking employment, by far the most common MO in South and Central Asia-connected reporting, also experienced a huge increase overall, increasing by more than a factor of five in number of reported cases, increasing from three to 14 percent of the total, and becoming the fourth most reported MO. Attempted acquisition of technology and request for information both claimed reduced shares, although they remained the third and fifth most reported MOs, and combined they still accounted for over a quarter of the total. Solicitation or marketing services remained stable, accounting for nine percent of the total both years.

The dip in cyber reporting linked to East Asia and the Pacific also affected collector affiliation data. DSS attributes a high percentage of East Asia and the Pacific SNA to the government affiliation. In overall FY13 data, government was the only affiliation that declined in number of reported cases, its share of the total fell from 25 to 17 percent, and it fell from the second to the fourth position. Commercial remained the most cited affiliation, accounting for 27 percent of the total, down slightly from the year before. The government-affiliated and individual categories rose in share, to 27 and 19 percent, respectively, taking over the second and third positions. This reflected the large increase in reported academic solicitation by individuals linked to government-related institutions or seemingly acting on their own.

DSS' shift in technology categorization schemes (discussed in the Background section) introduced some volatility to that aspect of the data. Nonetheless, electronics was the top targeted technology sector at eight percent, declining from 11 percent in FY12. The command, control, communication, and computers (C4) (five percent) and software (three percent) sectors—last year making up the top-ranked information systems category—combined to account for eight percent of the FY13 total, down from 11 percent in FY12. Similarly, the now-separate radars, optics, sensors (acoustic), and lasers sectors together accounted for seven percent, compared to their ten percent as one category last year.

Aeronautic systems rose from the fourth to the third most targeted technology sector in industry reporting, although it declined in both number of reported cases and share of the total. In contrast, marine systems increased almost 50 percent in number of reported cases and rose from the eighth most targeted sector to fourth.

The remaining reported collection efforts targeted technologies ranging widely over nineteen additional sectors. While none of these individually accounted for more than three percent, DSS considered the positioning, navigation, and time category to be especially significant. The special focus area of this publication addresses inertial navigation systems, a subset of that category.

Figure 3: Fiscal Year 2013 Regional Targeting Trends

Note: Categories in the legends are listed in order of prevalence based on overall fiscal year 2013 reporting.

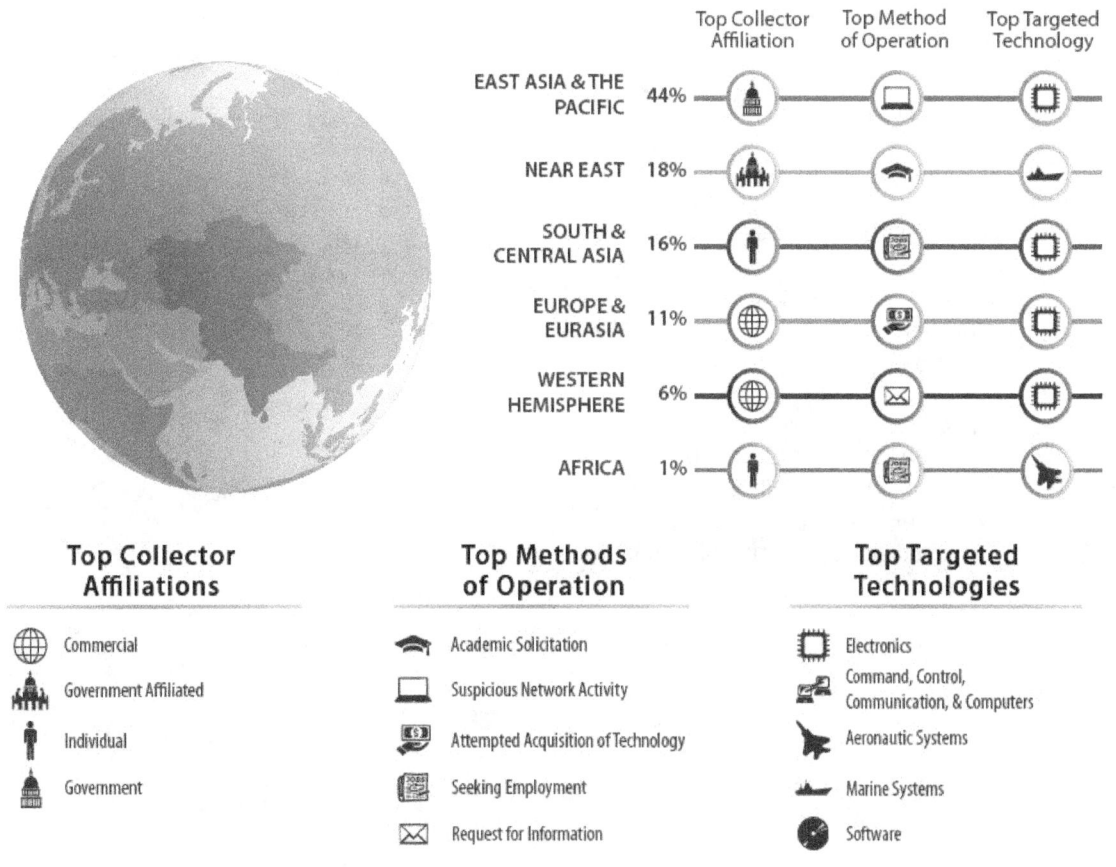

KEY POINTS
Based on FY13 industry reporting.

East Asia and the Pacific

• Continued to be the most prolific collector region, accounting for 44 percent of all reports

• While SNA remained the region's most common MO, there was a significant drop from FY12, whereas there was a substantial increase in academic solicitation

• Emphasized electronics, C4, aeronautic systems, and marine systems technologies

• Showed sustained interest in building business relationships with cleared industry

Near East

• Leveraged its network of intermediaries, procurement agents, brokers, and front companies, as well as access to cleared contractor facilities and personnel where possible

• Practiced academic solicitation as top MO (46 percent), but SNA capabilities improved

• Targeted research programs specializing in computational fluid dynamics—with applications to both marine and aeronautic systems—as well as electronics

• Government-affiliated collectors accounted for 44 percent of collection activities

South and Central Asia

• Amidst government intentions to produce more military technology indigenously, reported suspicious incidents linked to the region increased by two-thirds, making it the third most prolific

• Academic solicitation and seeking employment MOs accounted for nearly two-thirds of reported contacts, and individual affiliation more than doubled its share of the total

• Continued targeting electronics (especially enabling components) and C4; also focused on radars and nanotechnology

Europe and Eurasia

• Reported suspicious contacts increased over 50 percent

• Commercial entities were the most reported collectors, but individuals accounted for more than a quarter

• Attempted acquisition of technology remained the most reported MO, but seeking employment increased ten-fold and became the next most common

During World War II, German V2 ballistic missile scientists developed the first inertial navigation system (INS).[1] U.S. development of INS began in the late 1940s and early 1950s. The Massachusetts Institute of Technology Instrumentation Laboratory, Northrop, and Autonetics worked under U.S. Air Force sponsorship. This effort led to inertial guidance systems for ballistic missiles, both land- and sea-launched. The 1960s brought the Space Age and the advance of inertial navigation in the Apollo program. During this time, INS were also developed for military and commercial aircraft.[2] Since then, an INS has become standard equipment in ballistic missiles, aircraft, naval vessels, and spacecraft.

OVERVIEW

Foreign entities' interest in inertial navigation system (INS) technology has risen over the past several years, as reflected in industry reporting culminating in fiscal year 2013 (FY13). Reported targeting of INS technology rose nearly 60 percent from FY12. The Defense Security Service (DSS) produced this special focus area to alert cleared industry to the increasing foreign threat to INS technology and facilitate the implementation of mitigation strategies to counter that threat.

DSS analysis of industry reporting revealed that collectors demonstrating interest in advanced U.S. INS technology were linked to countries with INS production and manufacturing capabilities that could support the modernization of indigenous military inertial navigation. However, with few exceptions, those infrastructures still trailed those of the United States in producing key INS components that were both advanced and reliable.

An INS is comprised of an inertial measurement unit (IMU) and a computer. The IMU typically consists of gyroscopes and accelerometers. Gyroscopes measure the angular rate of change while accelerometers measure linear acceleration, both with respect to inertial space. An IMU may incorporate other motion-sensing devices, such as a magnetometer, which helps correct for drift. The computer takes the outputs from the IMU and calculates the vehicle's position, orientation, and velocity in reference to its starting point.

A human operator, a global positioning system (GPS), or some other external input provides an INS with its initial position and velocity. Thereafter an INS can compute the position and velocity of a moving vehicle by integrating information received from its motion sensors—the IMU components. The advantage of an INS is that after initialization it requires no external references to determine its position, orientation, or velocity. However, it can use external inputs, such as a GPS reading or star shot, to self-correct the system if the vehicle begins to drift during operation.

Many types of vehicles incorporate INS, including aircraft, submarines, spacecraft, and guided missiles. The cost and complexity of INS place constraints on the environments in which they are practical for use. Systems that employ mechanical gyroscopes and accelerometers typically cost less but are less accurate in determining orientation and speed. In contrast, nonmechanical laser gyroscopes offer more precise targeting, but are more expensive to research, develop, and manufacture. The main difference between the two is the greater rate of error for mechanical systems than for laser systems.

Many countries can manufacture mechanical gyroscopes, but very few can produce higher-end laser gyroscopes. Mechanical gyroscopes might be the preferred option for a country that relies on weapons systems employing numerous, fairly inaccurate missiles, such as Scuds. This is because the system is lost once used, so incorporating expensive INS is not cost-effective. Modern, nonmechanical gyroscopes might be used in military systems that require greater accuracy, such as aircraft, spacecraft, naval vessels, precision missiles, and unmanned aerial vehicles (UAVs). However, recent research and production advances in INS manufactured via microelectromechanical systems (MEMS)[i] have shown some promise for providing, almost as cheaply as pre-existing mechanical systems, a navigation capability as accurate as nonmechanical systems.

Within the 60 percent increase from FY12 to FY13 in industry reports of foreign collection attempts directed at cleared contractors that research, design, manufacture, and integrate INS technology, East Asia and the Pacific collectors were the most pervasive.

---

i MEMS – Miniaturized mechanical and electromechanical elements that are made using the techniques of advanced microfabrication. In the field of inertial navigation, companies are using MEMS technology to manufacture low-cost accelerometers and gyroscopes.

Their attempts accounted for 74 percent of all FY13 reporting from cleared industry. Collectors linked to the Near East, Europe and Eurasia, and South and Central Asia completed the top four in reported collection attempts against INS technology, with collectors associated with those four regions accounting for 96 percent of the FY13 total.

Although entities targeting U.S. INS technology linked to the second most active region, the Near East, accounted for only nine percent of cleared industry submissions, Intelligence Community (IC) reporting indicates the region maintains a persistent interest in this technology.

Based on industry reporting, foreign entities relied on five methods of operation (MOs) when targeting U.S. INS technology designers, engineers, manufacturers, and integrators: attempted acquisition of technology (AAT), academic solicitation, solicitation or marketing services, request for information (RFI), and foreign visit. These MOs accounted for 95 percent of FY13 reported collection attempts against INS-related technology, with AAT accounting for 37 percent by itself.

REGIONS TARGETING INS TECHNOLOGY

East Asia and the Pacific

East Asia and the Pacific entities accounted for 74 percent of FY13 industry reporting targeting INS technology. They sought information on all aspects of INS: gyroscopes, accelerometers, and MEMS sensing technology. Based on industry reporting, ring laser gyroscopes (RLGs)[ii] and fiber optic gyroscopes (FOGs)[iii] were the most sought-after INS technologies.

The region encompasses a robust missile guidance and control (G&C) infrastructure that includes inertial navigation test equipment for design, simulation, modeling, and data processing. It also has sufficient ability to design and manufacture strapdown INS[iv] that use dynamically tuned gyroscopes (DTGs) and quartz flexure accelerometers (QFAs).[3]

East Asia and the Pacific researchers have made progress in modernizing the region's INS technology. They have exhibited their work aimed at increasing the precision of their differential RLGs at international symposia. Research goals included more accurate position-determination capability for navigation technology incorporated into torpedoes, submarines, aircraft, and satellites. The researchers' work received sponsorship from military and other government institutions.

Because RLGs' high slew rate capability, scale factor stability, and ruggedness make them well-suited for strapdown INS, they are an attractive option for tactical missile system navigation. In addition, compared to mechanical systems' analog output, RLGs' digital output makes them an excellent choice for integration into satellite navigation capability. However, analysis of East Asia and the Pacific progress in developing modern, more accurate INS components indicates the region's accomplishments in this area remain several years behind those of the United States and Europe and Eurasia.

East Asia and the Pacific's lack of infrastructure to advance the development of modern INS components have impacts on commercial and military aviation industries. An East Asia and the Pacific institutional representative at a 2013 satellite

---

ii Ring laser gyroscope – A nonmechanical gyroscope that contains two counter-rotating beams channeled to a photo detector. If a vehicle is not rotating, the beams remain in phase. If the vehicle is rotating, one beam continuously changes phase with respect to the other. A diode translates the moving interference pattern into digital pulses representing the angle of rotation.

iii Fiber optic gyroscope – A nonmechanical gyroscope with the same operating principles as an RLG, but incorporating fiber optics to channel light.

iv A strapdown INS is rigidly attached to the body of the host vehicle, as opposed to a platform INS, which is mechanically separated from the host vehicle. The potential advantages of a strapdown INS are lower cost, reduced size, and greater reliability. Therefore, small, accurate, and lightweight strapdown INS may be fitted to small guided missiles.

navigation conference admitted that current INS technology available to the region was inadequate to support UAV navigation over long periods. To address this shortfall, researchers were working to integrate optical gyroscope INS, both RLGs and FOGs, into UAVs.

Another East Asia and the Pacific institution is researching FOG technology and has requested information from cleared industry. In April 2012, a cleared contractor employee received an RFI regarding dual-use FOG technology from a professor at an East Asia and the Pacific engineering university. Later, the professor approached the cleared contractor's display booth at a navigation conference in the United States and made several inquiries to the contractor's representatives regarding FOG and MEMS IMUs. The cleared contractor's booth neither displayed nor contained any literature pertaining to FOG technology.

The FOG IMU is a multi-use system, having military as well as commercial and scientific applications. It is an expensive, high-performance device that could

be installed on a variety of satellites. IC assessments indicate that any FOG technology that became available to the East Asia and the Pacific institutions in question would likely replace dynamically tuned gyroscopes in commercial applications and might be used in short-range ballistic missile guidance systems.

As an added concern, other IC assessments deem the world's most significant proliferators of modern missile inertial guidance components and their production infrastructures to reside in East Asia and the Pacific. These analyses have identified strong connections between these East Asia and the Pacific IMU development programs and similar inertial technology development programs in the Near East and South and Central Asia.

Analyst Comment: East Asia and the Pacific collectors likely are attempting to illicitly acquire advanced INS information and technology to support indigenous development of modernized nonmechanical IMUs. Current models are deployed as strapdown IMUs for ballistic missiles and some

Figure 4: Top Four Collector Regions

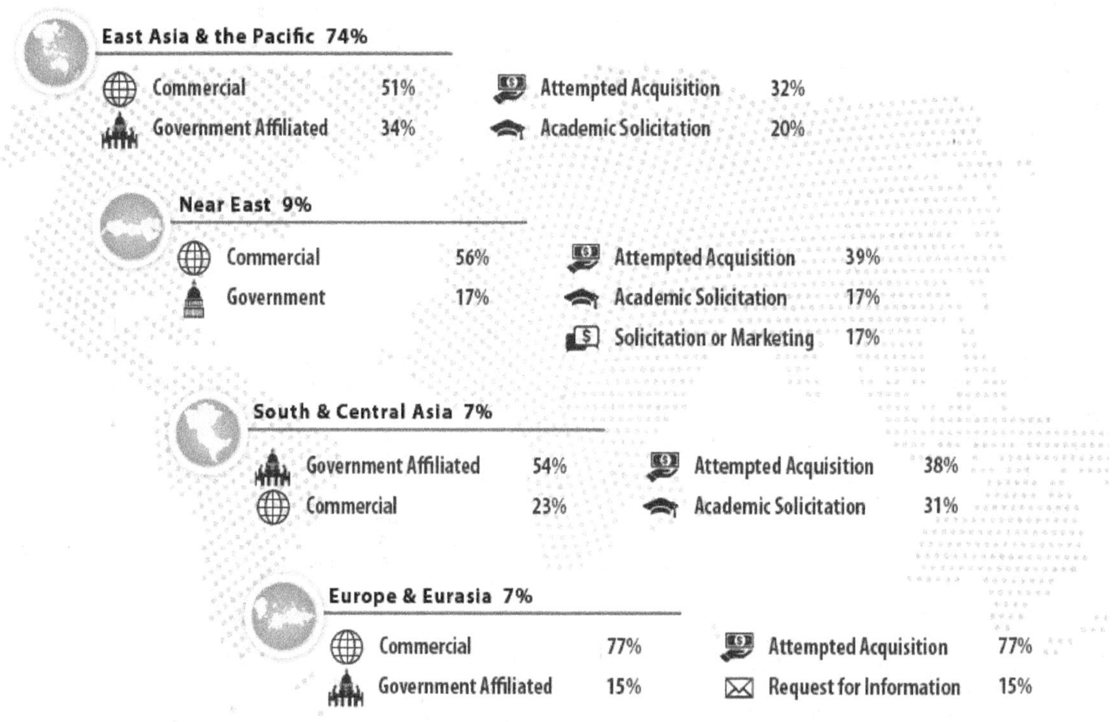

East Asia & the Pacific 74%

Commercial	51%		Attempted Acquisition	32%
Government Affiliated	34%		Academic Solicitation	20%

Near East 9%

Commercial	56%		Attempted Acquisition	39%
Government	17%		Academic Solicitation	17%
			Solicitation or Marketing	17%

South & Central Asia 7%

Government Affiliated	54%		Attempted Acquisition	38%
Commercial	23%		Academic Solicitation	31%

Europe & Eurasia 7%

Commercial	77%		Attempted Acquisition	77%
Government Affiliated	15%		Request for Information	15%

UAVs. East Asia and the Pacific entities probably intend to close the technological gap with Western countries engaged in developing advanced IMUs, likely leading to UAVs with less drift error and more accurate short- and intermediate-range and intercontinental ballistic missiles. (Confidence Level: Moderate)

East Asia and the Pacific's capability to exploit any illegally acquired INS technology likely puts U.S. INS technology at risk for proliferation to countries of concern in other regions. (Confidence Level: Moderate)

The Near East

Accounting for nine percent of FY13 industry submissions, Near East collectors were second in the number of reported attempts to collect U.S. INS information and technology. Based on FY13 data, Near East collectors primarily relied on AAT and focused their collection efforts on accelerometers.

The Near East contains some of the world leaders in the development of IMU and INS technology for military and commercial applications. Areas of high expertise include inertial navigation and gyro-stabilized electro-optical systems technology. Combat-proven products based on cutting-edge technologies find airborne, space, land, and marine applications worldwide.

Despite these regional capabilities, on multiple occasions Near East commercial companies or government-affiliated entities have contacted U.S. cleared industry and requested IMU technology. In March 2013, a cleared contractor received a request for quotation for 52 accelerometers. A Near East company made the request on behalf of a regional end user. However, previous IC reporting detailed an East Asia and the Pacific entity procuring military-grade technology from the Near East company, falsely claiming another Near East company as the end user. At least two of the components transferred in that transaction may have been of U.S. origin, which would have violated export controls.

IC assessments have found that Near East governments with access to U.S. technology information show a low willingness to safeguard it against exploitation or diversion, and have a history of transferring sensitive U.S. military technology and other U S.-derived technology to countries of U.S. concern without permission. Near East

entities persistently target Department of Defense technology, both to upgrade indigenous systems and to compete in the global marketplace. IC analyses deem it likely that, in the long term, Near East regimes will not protect any U.S. information and technology they obtain, will probably incorporate the technology into indigenous systems, and will possibly export the information and technology to other countries, including in other regions.

Analyst Comment: Despite the region's own capabilities for the development and manufacturing of INS technology, any acquisition of U.S. INS technology by Near East commercial firms likely puts the U.S. inertial measurement technology at risk. Near East commercial entities are probably seeking to acquire and exploit U.S. INS technology in order to upgrade systems for sale to their own governments or foreign countries. In addition, Near East firms probably intend to act as technology brokers, transferring any U.S. INS technology acquired from cleared industry to other countries, including countries of concern. (Confidence Level: Moderate)

South and Central Asia

In FY12, South and Central Asia-connected entities' share of industry reporting of targeting attempts against U.S. INS technology represented seven percent of the total. Reported collection efforts attributed to South and Central Asia entities primarily focused on MEMS and relied on academic solicitation.

According to IC reporting, South and Central Asia's most pressing technology targets are missile G&C and the associated electrical subsystems relating to various conformations of INS, IMU, and gyroscope and accelerometer technologies. Furthermore, the region's production capability is limited to low-quality gyroscopes and accelerometers, which are not suitable for high-precision short-range ballistic missile G&C applications. However, IC assessments find that, in contrast to some other regions, South and Central Asia possesses an advanced electronics industry that can meet some regional military INS demands.

Improvements in South and Central Asia production capabilities, especially regarding gyroscopes and accelerometers, will allow the region to minimize dependence on acquisition of foreign INS technology to support indigenous research and development (R&D). As early as 2004, regional militaries began updating inertial navigation components in acquired foreign ground warfare systems, replacing foreign ground-navigation devices with indigenous ones.

However, recent IC reporting indicates that South and Central Asia producers' advances in INS may not be able to provide all the advanced INS technology the region's militaries seek. To fill the gap, some regional actors seek outside assistance. There is a long history of military technology cooperation between South and Central Asia and European military industrial bases. In November 2013, a European delegation met with a high-ranking South and Central Asia government official to discuss a number of defense-related issues, including INS, missile technology transfer, and combat aircraft.

Analyst Comment: DSS assesses that South and Central Asia likely encompasses an advanced capability and infrastructure for the R&D, manufacture, and production of modern RLGs, FOGs, and accelerometers, and that regional entities are probably attempting to further advance INS technology development. However, industry reporting shows South and Central Asia entities continuing to attempt to collect MEMS technology from cleared industry, predominantly using academic solicitation; considering that reporting, the region likely has research gaps its collectors are trying to fill. (Confidence Level: Moderate)

South and Central Asia R&D efforts would probably apply any knowledge acquired from U.S. universities specializing in MEMS research to current indigenous MEMS research. South and Central Asia collectors' intention would likely be to advance their countries' INS manufacturing and production capabilities so as to support military missile, aerospace, and ground warfare systems modernization and development. (Confidence Level: Moderate)

Europe and Eurasia

Also accounting for seven percent of FY13 industry submissions, Europe and Eurasia entities rounded out the top collectors cited in reported attempts to collect U.S. INS information and technology. Like

Near East collectors, Europe and Eurasia collectors primarily relied on AAT. Unlike the reported Near East attempts to collect accelerometer technology, no one INS technology stood out as the most reported Europe and Eurasia target.

According to IC reporting, over the decades Europe and Eurasia manufacturers have produced many different types of accelerometers, including pendulum accelerometers using metal, quartz, and silicon hinges. Recently, Europe and Eurasia producers have demonstrated growth in the development and manufacture of functional and effective INS for military and commercial applications. Some of these producers' successes in developing the region's indigenous infrastructure for manufacturing IMU components came from exploiting U.S. INS technology, both accelerometers and gyroscopes.

For example, Europe and Eurasia producers developed DTGs for stabilized platform IMU applications as early as the 1980s, but their quality improved substantially after Europe and Eurasia technicians reverse-engineered a U.S. company's DTGs for use in both platform and strapdown IMUs. There are indications that in the mid-to-late eighties technicians reverse-engineered the U.S. company's QFA for two different applications, including an IMU using an RLG. Europe and Eurasia's optics infrastructure also supports the R&D and manufacture of guidance and navigation devices for missiles and other systems, including aircraft and satellites.

Europe and Eurasia research continues, aimed at maturing INS development and manufacturing capabilities within the region and maintaining INS parity worldwide. Research into an indigenous ability to manufacture inexpensive MEMS technology dates back to 1999, and has included publication of scientific papers, commercial-academic cooperation, and prototype development.

A majority of the requests from Europe and Eurasia entities for U.S. INS technology consisted of AATs from commercial entities. IC assessments show that multiple Europe and Eurasia governments have a low willingness to safeguard any U.S. technology and information acquired. Other reports also posited that some Europe and Eurasia regimes covertly acquire U.S. and third-country technologies by using commercial organizations for intelligence cover, then exploit and reverse-engineer those technologies. In addition, some Europe and Eurasia countries have technology-sharing agreements with countries of concern outside the region, including some covering missile technology, specifically navigation and G&C.

Analyst Comment: Europe and Eurasia commercial entities are likely seeking U.S. INS technology for one or both of the following purposes. First, given the extensive Europe and Eurasia history of reverse-engineering U.S. technology, including gyroscopes and accelerometers, it is almost certain that Europe and Eurasia technicians would reverse-engineer any advanced U.S. gyroscopes and accelerometers acquired to support indigenous space, commercial aviation, and military aviation or missile programs. Second, Europe and Eurasia entities would likely take an approach similar to that of some Near East firms: for financial gain, they would act as technology brokers. Europe and Eurasia commercial companies would probably transfer U.S. INS technology to third countries, including countries of concern. The most likely would be those countries with which they already have technology-sharing and assistance agreements. (Confidence Level: Moderate)

AFFILIATIONS AND METHODS OF OPERATION

According to DSS analysis of FY13 industry reporting, the collectors most active in collection efforts aimed at U.S. INS technology were those associated with East Asia and the Pacific, accounting for 74 percent of the total. Among reported East Asia and the Pacific collectors, it was commercial entities that most frequently attempted to obtain unauthorized access to sensitive or classified U.S. INS information and technology, in 51 percent of cases. East Asia and the Pacific entities used a variety of MOs. They employed AAT the most, in 32 percent of reports, academic solicitation in 20 percent of reports, and solicitation or marketing services in 19 percent.

East Asia and the Pacific use of AAT generally consisted of entities from the region submitting email purchase orders and quote requests for U.S. INS technology to cleared contractors. Soliciting or marketing services usually consisted of submitting an email offer to a cleared contractor to act as distributor or representative for the cleared contractor's INS in the region. East Asia and the

Figure 5: Programs Incorporating Inertial Navigation Systems at Targeted Facilities

East Asia & the Pacific

Vertical take-off and landing unmanned aerial system, Exoatmospheric Kill Vehicle, Global Hawk, electrostatically supported gyroscopic navigator, Joint Precision and Landing System, Mark 6 Guidance System, Trident, DDG 1000, Littoral Combat Ship, Landing Helicopter Assault Ship, NAVSTAR

Near East

Guidance and navigation research, joint robotics programs, navigation identification systems, standard missile, NAVSTAR, Air Traffic Control Approach and Landing System, C-130J, KC-10S, C-5, OH-58D, H-1, MH-47, CH-47, MQ-9

South & Central Asia

INS research, Ohio replacement, basic R&D in warfare and undersea technology, Gray Eagle, Predator, Reaper, Unmanned Carrier Launch Surveillance and Strike, RoadHawk, F-35, Advanced Hypersonic Weapon-Technology Demonstrator

Europe & Eurasia

INS Research and Development, A-10, B-2 Spirit, E-2D Advanced Hawkeye, E-8 Joint STARS, EA-18G Growler, F/A-18E/F Super Hornet, CH-47, MH-60R, P-8A Poseidon, MQ-9 Reaper, M1 Abrams, OH-58 Kiowa Warrior, C-5, C-130J Hercules

Pacific entities used the low-cost, potentially high-gain RFI MO in attempts to elicit information from cleared contractors in a variety of forums.

Analyst Comment: Suspicious network activity (SNA) was absent from the top reported East Asia and the Pacific MOs targeting INS technology. However, foreign collectors routinely exploit the anonymity of cyberspace and the vast number of opportunities provided by billions of interconnected devices and the trust users place in their digital systems. The region's cyber actors are likely employing SNA but the approaches are going unrecognized or unreported. When cyber actors successfully conduct an exfiltration, the victimized cleared contractor may be unable to identify what INS data, if any, was compromised. (Confidence Level: Moderate)

Reported Near East collection efforts relied primarily (56 percent) on commercial entities when attempting to acquire sensitive or classified INS information and technology. At 39 percent, AAT was the MO industry most frequently reported. Commercial entities emailed price quotes directly to cleared contractors that develop or manufacture

INS technology. These commercial entities often requested a specific number of IMUs, and at times attempted to obfuscate the end user of the INS.

The predominant South and Central Asia entities attempting to collect INS information and technology from cleared industry were government-affiliated, accounting for 54 percent. South and Central Asia entities practiced AAT in 38 percent of their reported attempts and academic solicitation in 31 percent. The students and academics in the latter cases sought research positions, primarily at facilities researching MEMS for missiles and aeronautics. South and Central Asia applicants highlighted advanced research experience in their résumés and curricula vitae.

Commercial entities accounted for 77 percent of the reported attempts to acquire sensitive or classified INS information and technology originating from Europe and Eurasia. Entities from this region most commonly used AAT, identified in 77 percent of the incidents. Request for information was the second most common MO, reported in 15 percent of incidents. The IMUs targeted, if acquired, would be suitable for incorporation into systems produced for domestic and foreign markets and would contribute to maintaining parity worldwide.

OUTLOOK

Reporting from industry confirms that advanced U.S. INS technology is of interest to entities in several regions. Technology collectors from these regions and of various affiliations will likely continue to use a variety of MOs to attempt to obtain unauthorized access to sensitive or classified U.S. INS technology. Cleared industry will likely continue to receive requests for INS technology as Western countries continue to develop technology offering greater accuracy yet lighter weight. (Confidence Level: High)

DSS assesses that the attempts of East Asia and the Pacific agents to collect INS technology will likely continue to dominate reporting in the future. (Confidence Level: High)

East Asia and the Pacific entities will almost certainly continue to use solicitation or marketing services in their attempts to entice U.S. INS manufacturers to sell their products to a large and growing regional aviation industry. They will also likely attempt to acquire advanced INS technology to replace antiquated mechanical and strapdown inertial navigation systems for missiles and military aircraft. In addition, given regional cyber actors' prolific use of SNA in general, they will almost certainly use this MO to supplement other collection methods in their targeting of U.S. INS technology. (Confidence Level: High)

Figure 6: Collector Affiliations

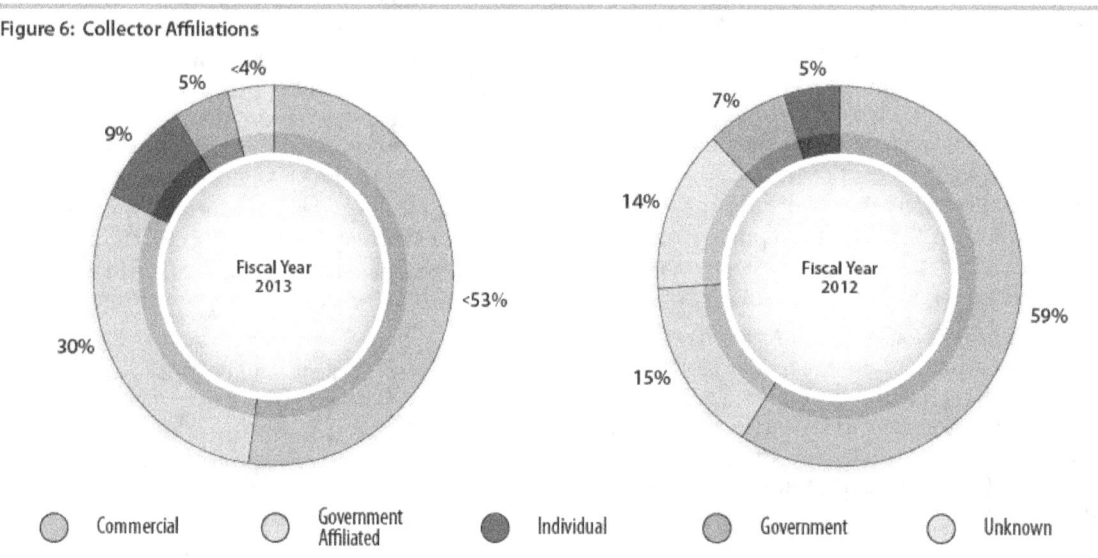

Although East Asia and the Pacific currently lacks the infrastructure for indigenous efforts to lead the development of FOG technology, there is an even chance that gaining access to U S. and Western university and cleared contractor research and technologies would allow East Asia and the Pacific to become a significant developer of FOG technology in the future. (Confidence Level: Moderate)

Near East commercial entities will likely continue to attempt to collect more advanced U.S. INS technology and exploit it in order to upgrade systems for sale to their governments or foreign countries. In addition, commercial entities from the region will probably seek INS from cleared industry, then act as technology brokers by re-exporting any technology gained, including to countries of concern. (Confidence Level: High)

South and Central Asia collectors will almost certainly continue to most commonly use AAT. The INS technology they target most will likely be MEMS, which could provide them with more advanced yet inexpensive INS manufacturing and production capabilities. The South and Central Asia push to modernize guidance and navigation on all weapon systems—missile, air, and ground—has developed a technologically advanced infrastructure within the region that includes commercial companies

capable of meeting military requirements for RLGs and FOGs. As U.S. MEMS R&D advances, students and academics from the region who have affiliations with their governments will likely continue their attempts to become part of U.S. MEMS research programs. Their intention will probably be to support South and Central Asia regimes' INS modernization of their military programs to close the INS gap with the United States. (Confidence Level: High)

Using INS technology collection methods similar to those of some Near East commercial entities, Europe and Eurasia commercial entities will likely continue to attempt to acquire export-controlled INS technology and sensitive information related to U.S. G&C. Should Europe and Eurasia commercial organizations or government entities acquire any U S. INS technology, they will very likely first attempt to reverse-engineer it, then incorporate advanced INS features or design elements into indigenous products for commercial and military applications. In addition, because multiple regional export regimes provide little protection to U.S. technology, Europe and Eurasia entities that acquire the U.S. information or technology will almost certainly seek to sell it to third countries of concern for profit. (Confidence Level: High)

The INS and Outs of Gyros

While foreign collection attempts reported in FY13 targeted all key components of INS, advanced gyroscopes, specifically RLGs and FOGs, were a highly sought after U.S. INS technology. Most foreign requests for inertial navigation technology were for gyroscopes that have military and commercial applications. East Asia and the Pacific commercial entities were the most prolific in FY13 reporting in their attempts to collect gyroscope technology. The case study below illustrates these patterns.

East Asia and the Pacific commercial entities have a significant history of attempting to obtain U.S. INS technology. In April 2013, a U.S. district court sentenced a citizen of an East Asia and the Pacific country to prison followed by supervised release for his role in a conspiracy to illegally export defense articles from the United States to East Asia and the Pacific. In February 2010, this individual, representing an East Asia and the Pacific company, emailed a cleared contractor and attempted to purchase two to five dynamically tuned gyroscopes, which are controlled under the International Traffic in Arms Regulations (ITAR). The individual stated the gyroscopes would be used in the energy industry. He promised a follow-on order of 50-100 units if the cleared contractor's gyroscope passed testing for the alleged end use.

In July 2012, authorities arrested the individual as he attempted to enter the United States. The district court indictment against him included one count of conspiracy to export defense articles from the United States without a license or approval from the State Department. According to the indictment, these gyroscopes can be used in strapdown INS for tactical missile guidance and unmanned aerial systems. Upon investigation, Immigration and Customs Enforcement alleged that the individual instructed other individuals in the United States to obtain and export the gyroscopes to his country and allegedly sought to use a courier to smuggle the gyroscopes out. The resulting indictment alleged that the individual, acting on behalf of a client in his country, sought to purchase three gyroscopes from an individual in the United States as a prelude to future purchases of gyroscopes.

Representatives of the East Asia and the Pacific company continued in their attempts to collect ITAR-controlled U.S. INS technology from cleared industry. A FY13 request from the cleared contractor differed from the indicted individual's request in the type of INS technology sought. The indicted individual had requested an INS technology that the IC assessed to be an area of manufacturing strength for his country: dynamically tuned gyroscopes. Conversely, the technology in the FY13 request was a more advanced gyroscope, a FOG, allegedly for the same end use.

The Department of Commerce has notified cleared contractors that received technology requests from particular East Asia and the Pacific countries that shipments of U.S. technology to certain companies could be further diverted. Additional IC reporting has identified companies that transship electronic goods to other regions. Investigations indicate that the use of East Asia and the Pacific-based companies to procure export-controlled electronics, including gyroscopes, from cleared industry is a common MO for illicit Near East procurement networks.

For example, in November 2011, a Near East procurement entity sent emails to a U.S.-based business requesting technology that was export-controlled. The technology requested was identical to technology an East Asia and the Pacific entity had previously requested. The procurement agent used email addresses registered to a Near East military college and a commercial company.

Analyst Comment: DSS deems that East Asia and the Pacific commercial entities' continued collection efforts exemplify the unremitting approach collectors from the region are using to attempt to collect U.S. INS technology. When East Asia and the Pacific commercial entities promise follow-on orders of hundreds of laser gyroscopes, a military missile or aircraft customer is probably driving the high demand. (Confidence Level: High)

The IC assesses that the East Asia and the Pacific infrastructure for modernizing military weapon systems with more accurate and reliable INS technologies is strong but still behind that of the United States and other Western countries. If East Asia and the Pacific commercial entities acquired RLGs and FOGs, they would almost certainly make these technologies available for government organizations to exploit. This would likely propel indigenous development of advanced strapdown navigation systems for missiles and more accurate and reliable UAVs. (Confidence Level: High)

An added concern is Near East illicit procurement networks using East Asia and the Pacific-based commercial entities to attempt to acquire U.S. INS technology. Near East R&D would likely use any technology acquired through these East Asia and the Pacific companies, first to modernize INS for legacy military aircraft, then to advance strapdown inertial navigation capabilities for tactical ballistic missiles. (Confidence Level: Moderate)

EAST ASIA & THE PACIFIC

OVERVIEW

Based on fiscal year 2013 (FY13) cleared contractor reporting to the Defense Security Service (DSS), East Asia and the Pacific entities once again were by far the most active in foreign collection attempts to obtain unauthorized access to sensitive or classified information and technology resident in the U.S. cleared industrial base. Reports from cleared industry associated with East Asia and the Pacific increased by 15 percent over FY12. These reports' share of the FY13 total was more than twice as large as that of the next closest region, and the number of reports linked to East Asia and the Pacific was nearly as large as the combined total for the next three regions.

East Asia and the Pacific contains some of the most frequently reported attempted collectors of sensitive or classified information and technology resident in the U.S. cleared industrial base. A number of countries in the region are or aspire to be world economic powers, and collectively the region represents a growing economic force. East Asia and the Pacific constitutes a very competitive economic environment, especially with regard to defense technology.

There is competition within East Asia and the Pacific to be the dominant regional power as well. Numerous conflicting land and sea border claims exacerbate existing geopolitical frictions, adding fuel to these rivalries. Multiple states in this region perceive neighboring states to be hostile, and likely future opponents in armed conflicts. The situation motivates East Asia and the Pacific powers' efforts to practice anti-access and area denial (A2/AD) doctrines.

Therefore, even as much of the world and region suffer from an extended period of economic retrogression, within East Asia and the Pacific retrenchment in spending, including for the military, is not universal. Some regimes prioritize continued defense spending over other budget sectors. To the extent that support continues within East Asia and the Pacific for military modernization efforts, it commonly involves research and development (R&D) of indigenous systems in parallel with targeting aimed at obtaining foreign information and technology relating to similar systems from the West. Even when indigenous East Asia and the Pacific R&D and production capabilities become nearly self-sufficient in some areas, regional collection efforts tend to continue to attempt to supplement these capabilities by acquiring foreign information and technology.

The role of the United States and its military forces in the region is in flux, becoming more prominent in some areas and less in others. Within East Asia and the Pacific, some perceive the U.S. military as a likely ally, others as a potential enemy. In general, however, the United States is perceived as a possible model and a source of useful military technology and related information.

While perceived potential threats from enemies and rivals both inside and outside the region remain the primary factor driving East Asia and the Pacific collection strategies, a secondary motivating factor is the desire to export even more arms to the rest of the world. As a strategy to further this agenda, East Asia and the Pacific interests frequently emphasize acquisition of dual-use technologies and practice reverse-engineering of U.S. and other Western technologies acquired.

East Asia and the Pacific reliance on foreign technologies includes both component systems and enabling technologies. However, as in prior years, much of FY13 industry reporting related to East Asia and the Pacific reflected interest in obtaining subcomponents rather than entire systems. Such subcomponents, which included various types of electronics, could be used in a wide array of systems, leaving their intended end use unclear.

This is especially worrisome given the application, frequently intentional, of dual-use technology to both military and civilian industrial sectors. Such integration can grow civilian science and technology sectors—which often enjoy greater access to foreign technology—that will eventually benefit defense industries.

In FY13, industry reporting revealed continued East Asia and the Pacific interest in almost all technology areas of the Industrial Base Technology List (IBTL). As with the Militarily Critical Technologies List (MCTL) in FY12, FY13 targeting data showed East Asia and the Pacific entities primarily seeking technologies in the electronics sector, which accounted for six percent of industry reporting. However, this share of the total as well as the sector's number of reported cases experienced a notable decrease.

Electronics was followed by command, control, communication, and computer (C4) systems (accounting for five percent of FY13 industry reporting), aeronautic systems and marine systems (four percent apiece), and software and radars (three percent apiece). In previous years' reporting, C4 and software systems were combined in the information systems technology category, in second place on the MCTL at nine percent; despite being separated into different IBTL sectors in FY13, both remained in the top five. A significant change in industry reports of the technologies East Asia and the Pacific entities targeted was the decline in the space systems category. The number of reported cases decreased by 50 percent, and the sector fell from the fifth most cited to the eleventh.

The affiliations of East Asia and the Pacific collectors in FY13 industry reporting remained consistent in ranking from FY12. However, while government entities remained the most frequently cited collectors, the number of cases attributed to this affiliation experienced a drop of over 20 percent in number of cases and fell from accounting for 41 percent of the total in FY12 to 28 percent in FY13. The primary factor was a decrease in the number of reports of suspicious network activity (SNA) that DSS was able to connect to the region. While the commercial affiliation increased in number of reported cases and remained in second place, its

share of the total also declined slightly, to 27 percent. The other three affiliations all increased in both number and share.

Many of the cases attributed to the commercial affiliation correlated with the continued prevalence in industry reporting of the solicitation or marketing services method of operation (MO). Intelligence Community (IC) reporting assesses that in many cases East Asia and the Pacific commercial entities work in cooperation with government intelligence services, especially with regard to technology collection and defense sales. Some East Asia and the Pacific intelligence services use commercial entities to attempt to collect information on U.S. technology and programs.

Like the commercial affiliation, East Asia and the Pacific government-affiliated entities also accounted for 27 percent of relevant FY13 data, representing a rise of 83 percent in number of industry reports from FY12. The majority of government-affiliated attributions were related to the ever-increasing number of academic solicitations from East Asia and the Pacific applicants.

Despite the collection activity attributed to other MOs above, in FY13 industry reporting the preferred MO of East Asia and the Pacific entities remained SNA. It accounted for 30 percent of the total, down from 42 percent in FY12. Academic solicitation followed at 20 percent, then solicitation or marketing services and attempted acquisition of technology (AAT) at 14 and 12 percent, respectively. Request for information (RFI), foreign visit, and seeking employment were all in the single digits for share of industry reporting.

But in contrast to declines in the number of industry reports of SNA, AAT, and RFI, the academic solicitation, solicitation or marketing services, foreign visit, and seeking employment MOs all increased in numbers of cases—at significant rates ranging from 35 to 300 percent—and in share of the total as well. Thus, overall in FY13, the region's reported collection activity skewed from more direct toward less direct methods of attempting to obtain unauthorized access to sensitive or classified information and technology resident in the U.S. cleared industrial base.

COLLECTOR AFFILIATIONS

Government entities were again the most active collectors in FY13 East Asia and the Pacific-related industry reporting, at 28 percent of the total. However, the commercial and government-affiliated shares were nearly as high, at 27 percent apiece. Combined, government, commercial, and government-affiliated entities accounted for over 80 percent of reported FY13 East Asia and the Pacific collection attempts. The individual and unknown affiliations accounted for ten and eight percent of industry reporting, respectively. Both of these percentages represented increases over FY12, and the number of cases in each category went up approximately a third.

Industry reporting of government entities' attempts to obtain unauthorized access to sensitive or classified information and technology resident in the U.S. cleared industrial base made it the top affiliation in FY13. However, the category experienced a drop of over 20 percent in number of cases and fell from accounting for 41 percent of the total in FY12 to 28 percent in FY13. The primary factor was a decrease in the number of reports of SNA that DSS was able to connect to East Asia and the Pacific. DSS categorized a high proportion of East Asia and the Pacific SNA as having a government affiliation, reflecting IC assessments that a portion of East Asia and the

Pacific computer network exploitation (CNE) activity may consist of a collection of efforts tasked by a central decision-making body.

Commercial entities were the second most common collectors in FY13 industry reporting. The number of cases attributed to the commercial affiliation rose 14 percent, but the category's share of the total fell slightly. Most submissions reported East Asia and the Pacific companies offering manufacturing services to cleared contractors, or commercial conference organizers soliciting participation in or attendance at upcoming conferences hosted in the region. The majority of East Asia and the Pacific companies involved were electronics manufacturers soliciting cleared contractors to integrate their components into U.S. technological platforms or requesting to serve as overseas distributors for cleared contractor products in regional markets. According to IC analysis, when East Asia and the Pacific commercial organizations manufacture electronic components for non-East Asia and the Pacific companies, it provides them an entrance into the U S. supply chain.

Analyst Comment: Although it is probably the profit motive that drives East Asia and the Pacific entities' solicitation of cleared contractors, allowing their components into the U.S. supply chain likely risks the degradation of critical military systems. (Confidence Level: Moderate)

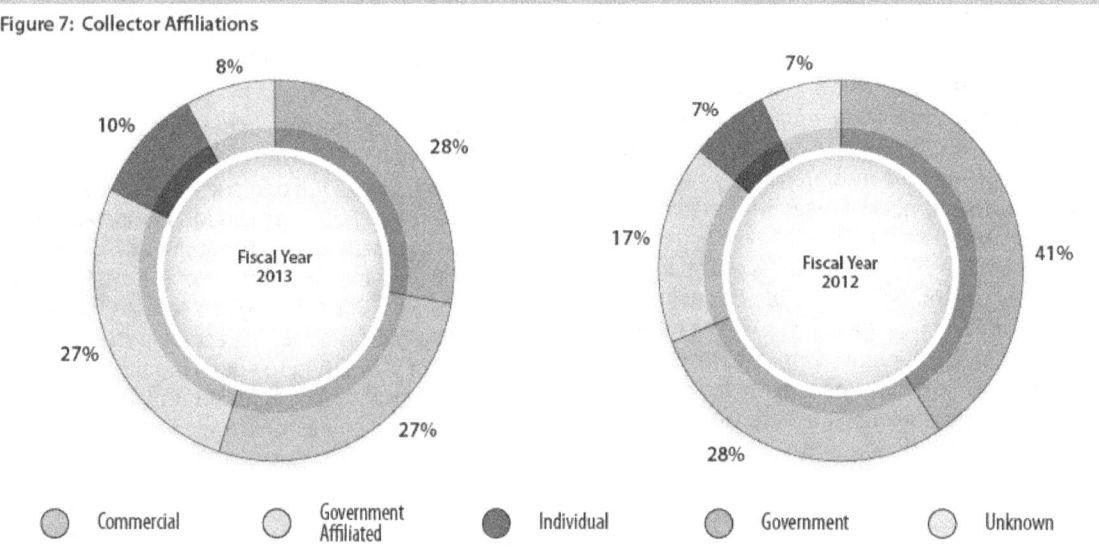

Figure 7: Collector Affiliations

Another pattern in cleared industry reporting involved various East Asia and the Pacific companies that requested to purchase U.S. information and technology but resisted providing end-user and end-use data. In some cases, IC reporting later identified these companies as procurement firms for East Asia and the Pacific militaries or state-owned research institutes (RIs) and factories. In February 2013, a representative of an East Asia and the Pacific company requested purchasing information for an advanced technology that has applications to satellites, ballistic missiles, and radars on behalf of an unidentified end user. IC reporting identified the company as having attempted to purchase sensitive, dual-use, and export-controlled components for East Asia and the Pacific government and government-affiliated recipients.

Analyst Comment: There is an even chance the companies omitted end-user information to attempt to obfuscate the identity of the ultimate consignee. Given that some dual-use items have applications to restricted end uses, East Asia and the Pacific procurement firms likely attempt to conceal military and R&D institutes as the recipients to convince cleared contractors the requested items will probably be used for civilian purposes. (Confidence Level: Moderate)

Compared to the commercial affiliation, the number of cases attributed to the government-affiliated category rose a much steeper 83 percent, and the latter's share increased by ten points to tie the commercial affiliation at 27 percent of the total. In FY13, reports citing the government-affiliated category primarily resulted from East Asia and the Pacific university researchers and students submitting résumés. Academics seeking internships and postdoctoral positions continuously submitted résumés for positions within U.S. defense programs, especially those specializing in nanotechnology, aeronautic systems, and space and marine systems.

In a significant majority of these cases, student and professorial applicants were associated with government-funded universities and institutes. Some of these institutions even sponsor, subsidize, and manage applicants' solicitations. IC reporting has identified some state-sponsored universities as providing training and R&D support to East Asia and the Pacific militaries. The IC assessed that the regimes involved use such enhanced academic programs to augment military R&D efforts by pursuing advanced research skills relating to critical technologies.

Analyst Comment: East Asia and the Pacific academics' access to the technological know-how resident in the U.S. cleared industrial base is grounds for concern. To the extent that they take any research results and knowledge acquired back to their home countries, it is likely such access ultimately supports their countries' military R&D. The sponsorship programs probably exert leverage on national students intended to engender targeting of cleared personnel and/or U.S. technology. (Confidence Level: Moderate)

In FY13, government-affiliated actors contacting cleared industry also included RIs and state-owned enterprises (SOEs). Although many of the RIs work on commercial or dual-use technologies, they are primary contributors to their countries' military R&D as well. The IC has assessed that any technology shared with them would probably be incorporated into the production lines of one or more East Asia and the Pacific militaries. FY13 cleared industry reporting supported this estimation by documenting various SOE representatives attempting to leverage partnerships with cleared industry to gain unauthorized access to U.S. information and technology.

Even though the number of reported cases that fell into the individual affiliation increased by 44 percent, its share increased only from seven to ten percent of the total. Similar to last year, industry submissions in the individual category largely consisted of reporting concerning East Asia and the Pacific applicants for jobs with cleared industry for whom research could not identify an affiliation with a government or commercial entity. This affiliation's share of the total remains relatively low because analysis usually allowed DSS to identify some connection between individuals and a commercial company or government or government-affiliated RI. Industry reported that the majority of individuals in this category were responding to cleared contractor vacancy announcements for jobs for which they were not eligible due to requirements for U.S. citizenship or a security clearance.

Analyst Comment: These employment solicitations were probably legitimate attempts to attain positions within cleared industry. However,

placement in sensitive positions likely provides individuals with opportunities to attempt to gain unauthorized access to U.S. information and technology. (Confidence Level: Moderate)

Many collection efforts originating in East Asia and the Pacific arise from some degree of coordination of effort between industry, government, and individuals. In numerous instances during FY13, multiple entities of different collector affiliations combined to attempt to collect against a single technology or cleared contractor. Between March 2012 and March 2013, a cleared contractor reported multiple attempts by various East Asia and the Pacific entities to obtain export-controlled radar technologies. The actors included commercial as well as government entities, both military and civilian. MOs included RFIs and exploitation of relationships, especially during cleared contractor visits to facilities in East Asia and the Pacific and foreign delegation visits to cleared contractor facilities in the United States.

Analyst Comment: East Asia and the Pacific collection attempts demonstrate a diverse and persistent approach, often involving multiple collector affiliations and MOs operating in concert. When one entity fails, a second entity, often of a different affiliation and using a different MO, engages the cleared contractor in pursuit of the same technology. FY13 East Asia and the Pacific collection efforts probably involved national government authorities setting general goals, followed by a pursuit of desired information and technology that involved industry, government, and academia operating under varying degrees of coordination. (Confidence Level: High)

METHODS OF OPERATION

The number of East Asia and the Pacific-connected SNA cases that industry reported declined 17 percent from FY12 to FY13. Nonetheless, based on industry and government partner reporting, SNA continued to be the foremost MO East Asia and the Pacific entities used in attempting to obtain unauthorized access to sensitive or classified information and technology resident in the U.S. cleared industrial base. This MO accounted for 30 percent of FY13 industry reporting linked to the region. East Asia and the Pacific actors remain the top cyber threat to cleared industry.

However, it was not just the SNA MO that experienced some volatility in its FY13 industry reporting statistics. Academic solicitation increased in number of reported cases by 93 percent, and went from the third most commonly cited MO in FY12 at 12 percent to the second most cited in FY13 at 20 percent. Solicitation or marketing services increased in number of cases by 48 percent and moved from fourth to third place, accounting for 14 percent of the FY13 total. AAT decreased nearly ten percent in number of cases, went from providing 16 to 12 percent of the total, and fell from second to fourth position. The foreign visit and seeking employment MOs maintained their rankings well down the list, but increased in number of reported cases by 35 and 300 percent, respectively. Only the RFI MO remained stable in its FY12 position, accounting for nine percent of the total.

The prominence of the SNA MO endured despite significant press coverage during FY13 detailing the results of Western research on and analysis of recent East Asia and the Pacific cyber activities. Reporting cataloged much of the infrastructure; command and control protocols; and tactics, techniques, and procedures (TTPs) East Asia and the Pacific cyber actors used. Industry submissions during the period immediately after the revelations decreased precipitately compared to the same period in 2012. U S. government partners also provided industry with additional technical indicators that can help identify suspicious activity, and continue to release indicators periodically.

But East Asia and the Pacific CNE tactics are constantly evolving as various sets of regional cyber actors learn from previous efforts, progress in their skills, and become ever more active. Increasingly, East Asia and the Pacific methodologies reflect technical cleverness and reliable intelligence practices. CNE originating in East Asia and the Pacific was responsible for many intrusions that cleared industry and government partners reported in FY13, some of which led to confirmed exfiltrations of data. East Asia and the Pacific CNE actors continue to defeat cleared industry network defenses. Fortunately, community reporting of SNA against cleared contractors provides indications and warnings to aid in mitigating and alerting others to planned or ongoing activity.

Analyst Comment: Open-source reporting very likely led to a decrease in industry-reported activity during FY13. East Asia and the Pacific CNE actors' becoming aware of government partners' release of indicators probably magnified the decrease. When East Asia and the Pacific CNE actors rebooted their CNE activity, they probably adjusted their methods and TTPs, thereafter employing a combination of old and new. This likely led to a further drop in reporting from cleared industry and government partners. (Confidence Level: High)

In FY13, a high percentage of East Asia and the Pacific cyber activity that cleared industry reported consisted of spear phishing. East Asia and the Pacific cyber actors' demonstrated preference for this widely used malware delivery method constitutes a persistent pattern. The remaining activity included website exploitation, compromised credentials, network scanning, brute-force attacks, distributed denial of service, and stack buffer overflow.

Analyst Comment: Spear phishing is a very adaptable delivery method by which cyber actors have achieved significant successes in soliciting access or information. Using it allows East Asia and the Pacific CNE actors to quickly change indicators, such as email addresses and links/attachments, with little cost or effort. East Asia and the Pacific CNE actors endeavor to remain a step ahead of countermeasures, and spear phishing will very likely remain their chosen tactic for delivering malware, gaining unauthorized access to sensitive or classified information, and making an initial contact to a cleared individual. (Confidence Level: Moderate)

It is very important for network defenders to determine the method used to deliver malware; knowing the delivery method allows network defenders to defeat some attacks. However, learning of successful compromises of a computer or network is crucial. In the aftermath of successful attacks, information concerning the compromises allows network defenders to conduct heuristic analyses and understand which technologies the attacks targeted and may have compromised.

In April 2013, East Asia and the Pacific cyber actors targeted more than a dozen different cleared industry members with an aerospace-themed spear-phishing email. It spoofed an executive of a

U S. company that provides products to the avionics industry. Most of the targeted cleared contractors were also in the avionics industry; several were in the aerospace industry; and others fulfill a support function. One cleared contractor's receipt of the spear-phishing email coincided with a genuine announcement about similar technology the spoofed and targeted companies had produced.

Details of the targeted email addresses provided evidence that the senders had pulled the addresses from a larger list. The email message asked recipients to "refer to the attachment" to retrieve details about the product. Embedded in the attachment was a malicious element in portable document format (i.e., PDF). The recipient's opening of the attachment opened the network to the sender and exposed company and government information and possibly other systems to compromise.

Analyst Comment: East Asia and the Pacific CNE actors very likely used the release of one cleared contractor's product to target not only that contractor but additional cleared industry members with a spear-phishing email designed to seem relevant and timely. The East Asia and the Pacific actors very likely conducted open-source research to obtain enough information about the cleared contractor and its product release to target individuals likely to open the attachment. Based on

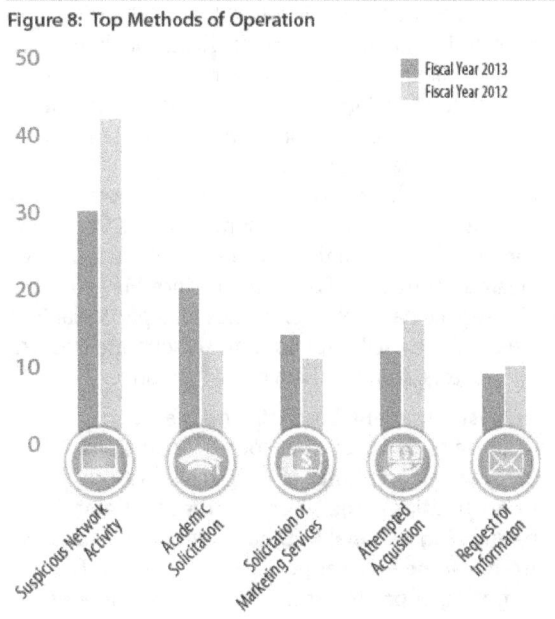

Figure 8: Top Methods of Operation

Fiscal Year 2013
Fiscal Year 2012

50
40
30
20
10
0

Suspicious Network Activity
Academic Solicitation
Solicitation or Marketing Services
Attempted Acquisition
Request for Information

the email addresses targeted, additional cleared contractors in the avionics industry probably received the spear-phishing email but it went unnoticed or unreported. (Confidence Level: High)

Within FY13 industry reporting linked to East Asia and the Pacific, multiple countries significantly, sometimes sharply, increased their generation of academic solicitation submissions to cleared contractors. Overall, the category almost doubled in number of cases and accounted for 20 percent of the total in FY13 industry reporting. This continued an upward trend noted in this publication over the last two years. Continued use of this MO provided evidence of a long-range, forward-thinking approach to promoting indigenous technology development. The related category of seeking employment increased even more in number of cases, almost tripling, and accounted for five percent of the FY13 total, up from one percent in FY12.

East Asia and the Pacific students and postdoctoral researchers often emailed a cleared contractor facility directly, seeking research positions or assistantships with specific professors. East Asia and the Pacific academics directed a much smaller number of requests to contractors not connected to universities, almost always inquiring about internships. While comparatively rare, academic solicitations also included requests to review and comment on research paper drafts.

In FY13, many academic solicitations that industry reported came from résumé services. For almost all the solicitations to U.S. universities, the service wrote the cover letter and résumé for the candidate; many cover letters contained very similar language. Cleared contractors often received applications from candidates whose areas of study were unrelated to a contractor's research. The résumé services frequently sent information on the same candidate to multiple cleared contractors. For example, since May 2012, one service has sent out résumés for a particular East Asia and the Pacific applicant to six separate cleared contractors, some of them more than once.

Analyst Comment: East Asia and the Pacific academics and their sponsoring institutions likely view acceptance to research and study opportunities in the United States as ultimately benefitting national R&D goals. Most solicitations from résumé services probably do not signify a targeting of particular technologies resident at the cleared contractors, but rather the breadth of information that East Asia and the Pacific entities seek. (Confidence Level: Moderate)

The bulk of East Asia and the Pacific solicitations sought postgraduate research positions related to a variety of research topics that included marine, aeronautic, and space system technologies, materials (raw and processed), and sensors (acoustic). The IC has concluded that access to cleared contractor information and technology related to these fields would further the long-term strategic defense goals of East Asia and the Pacific countries. These goals include developing indigenous capabilities that will reduce dependence on and/or vulnerability to U.S. technology and enabling such a country to dominate its region.

Analyst Comment: There is an even chance the increase in academic solicitations resulted from a similar intention among rival East Asia and the Pacific states to acquire expertise in R&D for their defense industries. In competition, each of these states probably sought thereby to further its goal of developing indigenous capabilities sufficiently to become the region's dominant power. Based on a significant increase in reported academic solicitations for a second consecutive year, DSS assesses there is an even chance East Asia and the Pacific entities attribute some success to this MO. (Confidence Level: Moderate)

Solicitation or marketing services was the next most common MO at 14 percent of FY13 industry reporting, up from 11 percent last year. In the majority of cases, cleared contractors reported East Asia and the Pacific commercial entities seeking to act as their sales agent or distributor in the region. Other contacts consisted of proposals to become the cleared contractor's supplier of certain components.

East Asia and the Pacific entities also commonly solicited via emailed conference invitations. Conference organizers enticed cleared contractor employees to attend a conference, serve on a panel, or deliver a presentation. Analysis showed that the vast majority of these invitations did not target subject matter experts' specialty areas directly— many individuals around the world received the invitations. However, DSS did note that several went to companies or individuals working in nanotechnology.

Analyst Comment: Whether invitations come from legitimate conference providers or at the direction of intelligence services, prominent use of this MO very likely signifies a concerted effort to have cleared contractor personnel attend conferences in the region. Such attendance almost certainly places cleared contractor personnel at risk for assessment and overt and/or covert collection activities. (Confidence Level: Moderate)

AAT and RFI were the fourth and fifth most commonly reported MOs, at 12 and nine percent of the FY13 total, respectively. In general, these industry reports resulted from East Asia and the Pacific commercial entities sending emails to cleared contractors seeking to purchase sensitive, classified, or otherwise controlled technology; asking for price quotes; or seeking information regarding a technology's capabilities. Requestors often provided little or no end-use information. In many of these instances, DSS subsequently connected what initially appeared to be a regular commercial request back to an East Asia and the Pacific military entity.

Analyst Comment: Most East Asia and the Pacific organizations almost certainly view AATs and RFIs as relatively simple ways to attempt to acquire desired information and technology. Cleared contractors often perceive these submissions as normal business behavior, so these MOs present little risk to East Asia and the Pacific entities. Yet while success via one of these requests may be rare, it can result in a significant payoff. (Confidence Level: Moderate)

In an example of AAT, in July 2013, a representative of an East Asia and the Pacific company emailed a cleared contractor seeking export-controlled signals intelligence (i.e., SIGINT) technology and associated hardware, but failed to reveal the prospective end user. The East Asia and the Pacific company has a history of providing sensitive U.S. technology to RIs in its country. When direct requests fail, the company often uses third-party intermediaries and provides false end-user information.

Analyst Comment: The East Asia and the Pacific company was likely attempting to obfuscate the identity of the ultimate end user of the requested technology. Based on the company's nefarious history and the sensitive nature of the technology in question, the company likely requested the technology on behalf of an RI. It is likely that any

technology acquired would ultimately benefit military R&D within East Asia and the Pacific. (Confidence Level: Moderate)

The bulk of RFI collection attempts consisted of email or web-card requests for price quotes, export requirements, or product specifications. In many instances, the requestor did not identify or was very vague about the end user and intended application.

In an example of RFI, in October 2012, an individual emailed a cleared contractor to request a price quote and delivery terms for a camera that was controlled under the International Traffic in Arms Regulations (ITAR). The individual did not identify his company or a specific end user, but requested expedited delivery of a sample camera for export to an East Asia and the Pacific country. The individual also stated that if the camera met the customer's needs, an order for 250 additional sets would follow. The individual sent a follow-up request in November 2012. Open-source searches revealed that the individual ostensibly operated at least four different named U.S. businesses, but research discovered no websites for them.

Analyst Comment: The individual's failure to identify the end user or his own company, his association with multiple companies with limited public footprint, and the purported urgency of his request all suggest he was not acting in good faith. There is an even chance the individual was acting on behalf of East Asia and the Pacific entities, and that he dangled the possibility of a lucrative deal to acquire a single camera, which end users would then have reverse-engineered without follow-on purchases. (Confidence Level: Moderate)

A number of AATs and RFIs made on behalf of East Asia and the Pacific organizations came through third countries. In FY13, DSS noted such requests coming from countries in East Asia and the Pacific, Europe and Eurasia, and North America. Some of these requests listed end-use information. In February 2013, a cleared contractor reported that an East Asia and the Pacific RI attempted to acquire, through a Europe and Eurasia company, short-wave infrared detectors for an unknown space-based system. East Asia and the Pacific entities have used the Europe and Eurasia country in question as a transshipment point in the past, as exemplified by a case in which a U.S. citizen pled guilty in July 2013 to illegally exporting high-tech materials from the

United States to East Asia and the Pacific through an unidentified company from the Europe and Eurasia country.

Analyst Comment: Regional collectors may use third-party intermediaries or proxies to lessen the apparent East Asia and the Pacific footprint with regard to certain purchases. There is also an even chance that such maneuvers reflect a lack of previous success in obtaining needed components via direct approaches to U.S. companies. (Confidence Level: Low)

The foreign visit MO accounted for six percent of FY13 industry reporting linked to East Asia and the Pacific. Countries from the region that succeed in establishing favorable defense trade relationships with the United States gain significant access to U.S. cleared industry. Associated collection entities often then attempt to capitalize on that access. In numerous reported FY13 cases, visiting delegations included several known or suspected embedded intelligence officers (IOs).

In one example from the spring of 2013, a cleared contractor hosted an East Asia and the Pacific delegation to cover an agreed-upon agenda that included unclassified discussions of several missile systems. The mixed military-civilian delegation of 13 East Asia and the Pacific nationals added two additional members at the last minute without providing their agency affiliations. Additionally, according to IC reporting, one member of the delegation had been a defense attaché at his country's embassy in Washington, D.C., as recently as November 2010, and during his stint conducted overt collection activities against technologies of interest to his government.

Analyst Comment: When delegation visits are an option for East Asia and the Pacific entities, those entities almost certainly attempt to exploit their access to U.S. cleared industry. Last-minute addition of personnel to delegations is typical of intelligence tradecraft, and in this case DSS assesses there is an even chance the roster changes inserted IOs to attempt to collect information regarding the missile systems. Even without embedded IOs, visiting delegations often attempt to steer discussions beyond the agreed-upon scope of subject matter. (Confidence Level: Moderate)

TARGETED TECHNOLOGIES

In FY13, East Asia and the Pacific entities targeted virtually every type of technology on the IBTL. Reported targeting efforts originating in the region spread broadly across the list, with individual sectors accounting for percentages ranging from six down to one percent. The top targeted technology in industry reporting was electronics, followed in order by C4, aeronautic systems, and marine systems.

However, analysis identified no IBTL technology in 23 percent of cases, and the targeted technology remained unknown in 31 percent more. DSS' identification of top targeted technologies is based on activity reports from cleared industry and other government partner reporting. Given that SNA was the most common MO in FY13 industry reporting linked to East Asia and the Pacific and that a large portion of cyber incident reports—nearly three-quarters of the total—could not specify a targeted technology, the few remaining cases that identify technologies can have a disproportionate effect on the data.

In cyber cases reported to DSS from cleared industry and other government partners, analysis can typically specify the targeted program or technology only when the victim identifies exfiltrated data or information or the CNE actors target a specific individual. In several reported FY13 intrusions, the cleared contractor was only able to determine the host computer or drives compromised. In some cases in which exfiltration could not be confirmed, cleared contractors assumed the intrusion had compromised all documents and files on the system or server.

There were additional impacts on the technology rankings from DSS' shift in categorization systems from the MCTL to the IBTL. The new IBTL broke down information systems, FY12's second most cited MCTL sector at nine percent, into C4 and software, which accounted for five and three percent, respectively, in FY13. Similarly, the component parts of FY12's third most cited MCTL category, lasers, optics, and sensors at eight percent, contributed seven percent under the IBTL in FY13, including radars as well. Most remaining technology sectors accounted for two percent apiece or less. This included positioning, navigation, and time (PNT), into which this publication's special focus area falls.

Analyst Comment: A shift in targeted technologies may have followed the publicity about East Asia and the Pacific CNE and the U.S. government's release of cyber indicators. Those events probably triggered a change in East Asia and the Pacific cyber actors' TTPs, creating a temporary blind spot for the computer network defense community and the IC. (Confidence Level: Low)

As with SNA, East Asia and the Pacific résumé submissions also seldom identified specific U.S. technologies. This forced DSS to categorize the majority of these reported incidents as no IBTL or unknown. When analysis could identify a technology in employment applications and academic solicitations, it frequently linked them to programs focused on nanotechnology, materials (raw and processed), and marine systems.

While DSS analysis of FY13 cleared industry reports attributed only two percent to the space systems sector of the IBTL, industry reporting on East Asia and the Pacific did reveal a general increase in interest in space-related technologies. Military strategists across the globe, including in East Asia and the Pacific, regard as central to success in modern warfare the ability to use space oneself and deny adversaries access to it. For example, space operations can serve a key role in enabling A2/AD efforts, which multiple East Asia and the Pacific states would aim to conduct in the event of a conflict within the region, in part to disrupt any U.S. intervention. East Asia and the Pacific states continue to attempt to fill technology gaps in their developing and modernizing space programs by acquiring U.S. technology. This general concentration on space applications affects several other technology sectors.

Despite a decrease of over 30 percent in the electronics sector's number of reported cases and from ten to six percent in its share of total industry reporting, it remained the most targeted single technology sector in FY13 industry reporting related to East Asia and the Pacific. Reported requests often concerned space-qualified radiation-hardened (rad-hard) integrated circuits, special monolithic microwave integrated circuits (MMICs), semiconductors, or power amplifiers. East Asia and the Pacific collectors' concerted efforts to obtain rad-hard integrated circuits provided a prime example of

their focus on obtaining technologies that contribute to space programs, since the region continues to struggle to produce these circuits indigenously.

For instance, a cleared contractor specializing in rad-hard integrated circuits received multiple requests from East Asia and the Pacific-based entities attempting to procure rad-hard components. DSS identified many of the East Asia and the Pacific companies as import/export firms that attempted to obscure the ultimate recipient. Notably, the multiple solicitations for rad-hard components were often consistent in specifications and quantities.

Analyst Comment: Given the similarities in the types and quantities of circuits that numerous East Asia and the Pacific entities requested, the requestors were almost certainly acting on behalf of a common intended end user. Given the lack of end-use information provided in conjunction with the need East Asia and the Pacific authorities perceive for rad-hard components for space applications, DSS assesses that East Asia and the Pacific regimes would very likely have used any circuits acquired to support their space-related R&D goals. (Confidence Level: High)

Cleared industry reported numerous requests for other space-qualified electronics such as the specialized MMICs, which are ITAR-controlled. A variety of civilian and military space systems incorporate these electronics, including missiles, traveling wave tube replacements, and satellites. East Asia and the Pacific RIs labor to develop and produce such specialized MMICs, but—as with rad-hard components—they have struggled to produce indigenous technology with quality comparable to

Figure 9: Top Targeted Technologies

Electronics 6%
Radiation-hardened integrated circuits; monolithic microwave integrated circuits; semiconductors; power amplifiers

Command, Control, Communication, & Computers 5%
Waveguides; airborne data acquisition systems; data links; man-portable satellite communications terminals

Aeronautic Systems 4%
Fighter aircraft; unmanned aerial vehicles

Marine Systems 4%
Autonomous underwater vehicles; academic programs (computational fluid dynamics)

U.S. variants. Cleared industry reporting documented that entities associated with East Asia and the Pacific RIs attempted to procure MMICs. For example, in August 2013, one such entity submitted a webcard request for the specialized MMICs. While the requesting company stated the application would serve base stations, it did not provide an end user.

Analyst Comment: The specialized MMICs in question would very likely fill technology gaps in East Asia and the Pacific space systems. Regional militaries would also likely integrate any such technology acquired into their R&D. Based on the requesting company's direct and indirect connections with an East Asia and the Pacific RI, the company was likely attempting to acquire the technology on the RI's behalf. (Confidence Level: Moderate)

In FY13, another commonly targeted technology category related to East Asia and the Pacific space system development was software, which accounted for three percent of industry reporting. Primarily these efforts consisted of academics from the region attempting to procure modeling and simulation (M&S) software. Space applications include satellite control, atmospheric modeling, and launch planning and analysis, while other applications include failure prediction in circuit boards.

While such software is often used for satellites, it can also be used for aircraft, missiles, and other exo-atmospheric vehicles. By enabling simulators that facilitate military training doctrine, M&S software helps improve warfighting capability within East Asia and the Pacific. However, in 2012 the IC assessed that East Asia and the Pacific M&S software was less sophisticated than Western versions. In FY13, cleared industry reporting documented that East Asia and the Pacific academics and individuals tied to regional militaries conducted various aggressive collection efforts against U.S.-developed M&S software.

For example, various attempts to download satellite-control software from a cleared contractor website emanated from East Asia and the Pacific. The efforts involved several different names, affiliations, and email addresses. IC reporting indicated that many of the Internet protocol addresses overlapped with proxy servers that East Asia and the Pacific cyber actors use to conduct defense-related open-source intelligence collection.

Analyst Comment: Given the associations with an East Asia and the Pacific military and the persistent and aggressive attempts to download the software program in question, DSS assesses that within East Asia and the Pacific there is very likely an ongoing collection requirement for space-related M&S software. (Confidence Level: Moderate)

Efforts within East Asia and the Pacific to modernize military technology, particularly satellite and naval programs that provide options against regional rivals, continue to be substantial. IC observers believe regional powers are actively pursuing an A2/AD capability to deter possible U.S. intervention in any regional conflict. Broader goals of modernization include enhancing naval capabilities to defend territorial claims in the South and East China Seas and combat U.S. influence in the Pacific.

Achieving space and naval superiority in conjunction with A2/AD capabilities requires upgrading C4 systems, the second most commonly targeted technology sector in FY13 industry reporting linked to East Asia and the Pacific, at five percent of the total. Achieving naval superiority, at least with regard to regional naval missions, requires effective ship-to-shore communication links, which today mostly rely on microwave technology.

The IC assesses that East Asia and the Pacific navies aim to install waveguides on their ships to support this communications technology. Besides deploying their current waveguides, East Asia and the Pacific collectors continued to request waveguides resident in the U.S. cleared industrial base. East Asia and the Pacific commercial entities targeted ITAR-controlled waveguides particularly aggressively, as evidenced by a cleared contractor that received numerous requests in quick succession for price quotes for waveguides. Almost all the requests were for identical components and quantities. The companies and other entities behind this procurement effort have a history of requesting export-controlled technologies while maintaining virtually no Internet presence.

East Asia and the Pacific collectors also sought other C4 components that would enhance battlefield communication, including airborne data acquisition systems, antennas, and connectors. In addition, they sought data links for unmanned ground and aerial systems and man-portable satellite communications terminals.

Analyst Comment: The effort numerous foreign companies exerted in attempts to acquire this communication equipment very likely signifies a military collection requirement intended to bolster East Asia and the Pacific C4 systems for naval communication. Collectors likely used companies with limited Internet footprints to obscure the end user and evade export controls. (Confidence Level: Moderate)

The airborne data acquisition systems mentioned above were a frequently targeted C4 technology. The East Asia and the Pacific region produces commercial aircraft. The commercial entities that do so often share facilities, personnel, and RIs with government entities while simultaneously maintaining relationships with numerous U.S. aviation companies for developmental work related to indigenous aircraft.

In 2012 reporting, the IC determined that the government entities involved leverage these joint business ventures to attempt to collect sensitive aerospace information and technology. Cleared industry reporting provided confirmation in FY13 in the form of solicitations received from East Asia and the Pacific entities that seemingly sought to leverage such relationships to circumvent export controls.

For example, in August 2013, an East Asia and the Pacific-based company contacted a cleared contractor requesting information regarding a proposal to provide data acquisition systems for an East Asia and the Pacific-based aircraft producer. Separate 2012 IC reporting stated that the initiating East Asia and the Pacific company was a suspected front company that the IC suspected of previously attempting to unlawfully acquire export-controlled technology from a separate U.S. cleared contractor.

Analyst Comment: The aircraft producer's established partnerships with cleared industry almost certainly provide avenues for attempted exploitation. The ultimate goal of these efforts was almost certainly to enhance its indigenous R&D efforts and, by extension, those of the government entities with which the producer is also associated. (Confidence Level: High)

After the broad technology categories of electronics and C4, East Asia and the Pacific entities targeted aeronautic systems next most often, in four percent of all reported suspicious contacts. Within aeronautic systems, collection efforts focused on fighter aircraft and unmanned aerial vehicles (UAVs). Typically, East Asia and the Pacific powers follow a two-pronged strategy of simultaneously pursuing the purchase of advanced aeronautic technology resident in the U S. cleared industrial base and the development and deployment of indigenous aircraft. However, in FY13, budget constraints and feasibility issues often delayed or prevented bringing one prong or the other to fruition.

Analyst Comment: If an East Asia and the Pacific power purchases a U.S. aircraft, the aircraft will almost certainly satisfy its operational expectations. However, if the country has an indigenous aircraft development program, it will likely create additional collection requirements for advanced aeronautic systems technologies. (Confidence Level: Moderate)

Regarding UAVs, East Asia and the Pacific entities also targeted various technologies in this area in FY13. Regional producers have worked on their own UAVs for years and anticipate deploying new models within two years, but they also seek high- and low-altitude, long-endurance models from the United States. IC assessments have determined that East Asia and the Pacific powers pose a medium collection threat to U.S. UAV technology, will likely attempt to reverse-engineer any advanced UAV technology they obtain, and may transfer this technology to foreign UAV customers.

While armament and survivability technologies constituted only the seventh most targeted category and accounted for only two percent of FY13 industry reporting, the sector's number of reported cases increased more than 40 percent. Specific technologies targeted included guns, missiles, rockets, and personal protective equipment. These technologies would be useful for responding to a conventional invasion.

Analyst Comment: Given the surge in reported collection attempts against armament and survivability technologies, East Asia and the Pacific states probably have shortfalls in critical munitions. Likely causes include insufficient budget allocations and production time limitations. East Asia and the Pacific states that perceive themselves to be endangered by regional defense threats are probably engaging in collection activity intended to address their shortages in order to meet those challenges. (Confidence Level: Moderate)

East Asia and the Pacific entities sought a variety of PNT technologies, with the sector accounting for two percent of FY13 industry reporting. Collectors paid particular attention to space-qualified accelerometers, guidance kits for munitions, and angular rate sensors. A majority of the collectors were commercial entities. It is particularly worrisome that East Asia and the Pacific entities have demonstrated a willingness and the ability to reverse-engineer PNT systems such as accelerometers and wireless global positioning systems (GPS). East Asia and the Pacific technicians reverse-engineered quartz flexure accelerometers from U.S. accelerometers that were designed for aviation, not space applications.

Analyst Comment: East Asia and the Pacific entities are almost certainly attempting to upgrade their accelerometers by acquiring U.S. space-qualified accelerometers, probably to meet space program demands. As they have in the past, East Asia and the Pacific entities would likely attempt to reverse-engineer any accelerometers procured. (Confidence Level: Moderate)

Other East Asia and the Pacific entities attempted to collect PNT technologies involving anti-jamming technologies for GPS. One attempt was against gyroscopes, a component of the subject of this publication's special focus area, inertial navigation systems. An East Asia and the Pacific national representing a defense company from his country requested information regarding a cleared contractor's lithium-niobate fiber optic gyroscope at a U.S. trade show. For further discussion of this technology category, see the special focus area section of this publication.

Analyst Comment: The single collection attempt against gyroscope technology was a relatively innocuous request made at a convention whose very purpose was to advertise technologies. Given the lack of additional requests relating to gyroscopes, East Asia and the Pacific entities very likely do not perceive the technology to be a requirement at this time. (Confidence Level: Moderate)

The East Asia and the Pacific practice of using the academic solicitation MO as part of long-term efforts to develop a knowledge base that can support development of emerging technologies was

especially apparent in relation to nanotechnology. While the technology sector as yet accounts for only two percent of industry reporting, it represents a cutting-edge field. East Asia and the Pacific regimes continue to demonstrate their willingness to expend considerable effort and resources to advance indigenous nanotechnology, especially that having potential military applications. In particular, East Asia and the Pacific universities place significant emphasis on research into nanotechnology. However, the weight of industry and intelligence reporting indicates that a large qualitative gap remains between indigenous and U S. nanotechnology, and East Asia and the Pacific regimes desire to avoid spending resources to "reinvent wheels" that others, including the United States, have already developed.

Cleared industry reporting in FY13 documented various approaches from East Asia and the Pacific academics soliciting postdoctoral or research positions relating to nanotechnology and materials (raw and processed). The materials field often supports nanotechnology development. Like nanotechnology, materials represented two percent of total FY13 industry reporting. The number of materials cases was slightly lower than in FY12 MCTL data; nanotechnology only became a separately tracked category with the switch to the IBTL in FY13.

OUTLOOK

Industry reporting on East Asia and the Pacific collection efforts has increased every year DSS has tracked statistics, and cleared industry will almost certainly continue to experience a growing number of suspicious contacts from East Asia and the Pacific entities. This region's collection attempts will almost certainly continue to represent the most prolific threat to cleared industry. (Confidence Level: High)

Multiple countries within East Asia and the Pacific are readjusting their strategies, policies, and military stances due to changes in the U.S. role in and approaches to the region. Resultant tensions, along with preexisting ones arising from long-standing territorial claim conflicts, border insecurities, and unfriendly neighbors, will very likely further motivate East Asia and the Pacific collector activity. As East Asia and the Pacific militaries continue their attempts to modernize, those efforts too will very likely continue to drive attempts to obtain unauthorized access to sensitive or classified U.S. information and technology. (Confidence Level: High)

DSS assesses that many East Asia and the Pacific technology collection efforts likely represent more or less coordinated national strategies that leverage multiple entity types. In FY13, government, commercial, and government-affiliated entities contributed almost equal proportions of industry data related to East Asia and the Pacific, and this relatively balanced, variably integrated approach will very likely continue. (Confidence Level: High)

SNA will very likely continue to be the top reported MO in reported targeting of cleared industry networks linked to East Asia and the Pacific. The FY13 open-source reporting on East Asia and the Pacific CNE, reinforced by government-provided indicators information, almost certainly had a significant impact on East Asia and the Pacific computer network operations (CNO). However, by June 2013, industry reporting reflected that East Asia and the Pacific cyber actors began to re-establish their foothold in CNO aimed at cleared industry. In FY14, East Asia and the Pacific entities will probably continue to regain their momentum in terms of successful intrusions into cleared industry networks, leading to further thefts of critical technology data. (Confidence Level: High)

East Asia and the Pacific CNE actors have achieved success in penetrating cleared industry networks using spear phishing, at little cost in effort and resources. Such success confirms that, year after year, the vector remains not only efficient but difficult to protect against. The vector East Asia and the Pacific cyber actors most commonly use in their attempts to compromise cleared industry networks will almost certainly continue to be spear-phishing emails containing malicious files or suspicious links. (Confidence Level: High)

In FY12, DSS noted an upturn in academic solicitation and the need for additional reporting to determine whether this constituted a trend or an anomaly. The two-year increase in academic solicitation likely means that East Asia and the Pacific entities experienced success with this MO. Especially when the "push" of state sponsorship is combined with the "pull" of U S. universities' reputations, East Asia and the Pacific university students and

professors will likely continue to apply for positions at these universities, and industry reports of academic solicitation linked to the region will very likely continue to increase. (Confidence Level: Moderate)

In conjunction with academic solicitations, there is an even chance nanotechnology will become a more frequently targeted technology sector due to increased funding for and emphasis on nanotechnology research within East Asia and the Pacific. Such an increase in reported targeting of nanotechnology would likely result primarily from postdoctoral researchers and students submitting résumés to U.S. university programs that specialize in the field. (Confidence Level: Moderate)

The solicitation or marketing services MO, primarily practiced by commercial entities, will very likely continue to appear frequently in industry reporting. East Asia and the Pacific entities will also almost certainly continue to make direct requests in the form of RFIs and AATs to attempt to procure sensitive or classified information and technology. As noted in the FY12 version of this publication, the continuation of a sizable volume of such requests likely means East Asia and the Pacific entities have had some measure of success from employing résumé submissions, solicitations, and direct requests. (Confidence Level: High)

In part to further ongoing military modernization efforts, East Asia and the Pacific entities will almost certainly continue to attempt to collect against a broad spectrum of sensitive, classified, and export-controlled technologies. In addition, wide-ranging and continuing efforts to enhance the region's naval and space programs will almost certainly continue to drive much of the collection activity against U.S. information and technology. (Confidence Level: High)

Given the geography of the East Asia and the Pacific region, regimes consider maritime power to be especially important, both as a defensive necessity and for its offensive potential. This will almost certainly continue to fuel East Asia and the Pacific collectors' attempts to steal data in related technology sectors, especially in marine systems subareas such as autonomous underwater vehicles. (Confidence Level: High)

Along with marine systems, electronics, C4, and aeronautic systems technologies will very likely continue among the four most targeted technology sectors in FY14 industry reporting and beyond. Technology in these areas is undergoing dynamic growth and transformation, and East Asia and the Pacific entities almost certainly recognize technology gaps they desire to close. (Confidence Level: High)

East Asia and the Pacific companies will likely continue to attempt to use partnerships with U.S. businesses to achieve military R&D goals and bolster indigenous production capabilities. East Asia and the Pacific entities will probably attempt to exploit partnerships such as those associated with commercial aviation programs to fill military aerospace technology gaps. This strategy will likely contribute to C4 and aeronautic systems remaining among the most commonly targeted technologies. (Confidence Level: Moderate)

Even if East Asia and the Pacific regimes directly acquire manned and unmanned aircraft from the United States, ongoing intentions to produce indigenous aircraft will likely maintain additional collection requirements for advanced aeronautic technologies. DSS assesses that aeronautic systems will likely remain a top-targeted technology category in FY14 reporting linked to East Asia and the Pacific. (Confidence Level: High)

Countries with weak export-control regimes place resident technology at risk of illicit transfer, and technologies that East Asia and the Pacific entities obtain are almost certainly at risk of diversion to third countries. Furthermore, to sustain the viability of their defense industries, East Asia and the Pacific states sometimes make technology transfer a part of foreign military sales. Therefore, any technology either shared with entities from these countries or obtained via unauthorized access is very likely at risk of further proliferation. (Confidence Level: High)

M&S= Meteorology & Science? Or Military & Space?

The following case study highlights various East Asia and the Pacific efforts to procure satellite control software.

A cleared contractor produces modeling and simulation (M&S) software that models spectral resolution and atmospheric correction. Beginning in August 2013, the contractor received four requests in one month for its software. One request originated from an East Asia and the Pacific national who was serving as a visiting scholar at an identified U.S. university and claimed to represent an East Asia and the Pacific meteorological institution.

The meteorological institution is subordinate to a quasi-governmental body that provides the military with meteorological observation and forecasting support. The IC assessed in 2012 that cyber actors from the same country have targeted U.S. National Oceanic and Atmospheric Administration meteorological satellites to support their country's weather agencies, and that the country's military relies on meteorological satellite images to plan military operations and conduct exercises.

In September 2013, the same cleared contractor received three additional requests for the same M&S software. The email address of the first of the requestors resolved to a company whose primary customers are an East Asia and the Pacific military and other government entities. The second request came from an East Asia and the Pacific software company, which failed to cite an end user. After the cleared contractor elected not to respond to these requests, a third regional company requested the software. Further analysis revealed that the third company serves as the second company's distributor within East Asia and the Pacific.

Analyst Comment: When used with satellites, M&S software provides critical information concerning all aspects of exo-atmospheric flight. As East Asia and the Pacific military space programs continue to expand, including by deploying more military satellites, East Asia and the Pacific collectors will likely continue to consider such software a critical requirement. (Confidence Level: Moderate)

Past DSS analysis of cleared industry reporting has frequently cited East Asia and the Pacific entities as requesting M&S software, particularly for satellite applications. Typically, East Asia and the Pacific procurement attempts have employed U.S.-based individuals and organizations, East Asia and the Pacific suppliers acting as intermediaries, and obfuscation of requestor identities. East Asia and the Pacific collectors are likely to continue using all these techniques as they seek to fill technology gaps within their M&S software capabilities. (Confidence Level: Moderate)

THE NEAR EAST

OVERVIEW

In fiscal year 2013 (FY13), the Near East continued to face various regional concerns and struggles: civil war in Syria, border issues, intraregional hostilities and rivalries, and internal conflicts. A single prolonged engagement or a combination of simultaneous conflicts could ultimately overwhelm any regional actor's defenses. Near East states perceive both conventional and asymmetrical threats, ranging from straightforward invasions to missile barrages, terrorist assaults, and cyber attacks.

Reflecting both existing and potential conflicts and varying national geostrategic priorities, FY13 Near East collection efforts, as reflected in cleared industry reporting, corroborated Intelligence Community (IC) assessments that Near East entities continued to seek a wide variety of military and dual-use technologies. They likely intended to employ any technologies gained to maintain internal security, monitor dangers at borders and beyond, provide active external defenses, and support indigenous defense industries.

Both industry and IC reporting indicated that Near East collectors continued to actively attempt to obtain unauthorized access to sensitive or classified information and technology resident in the U.S. cleared industrial base, whether in contravention to trade assistance agreements, U.S. law, or U.S. and international sanctions. As states struggled to maintain both current defense operations and developmental programs despite generally falling defense budgets, foreign entities linked to them continued to ply their collection methodologies.

These entities sought access to U.S. information and technology through networks of procurement agents, technology brokers, front companies, and intermediaries; through cooperative and joint ventures; via personal contact during foreign visits or at defense exhibitions; or by directly pursuing

acquisition of or information about defense technology. Use of email and cyber exploitation techniques was on the rise.

Based on industry reporting to the Defense Security Service (DSS) from FY13, entities linked to the Near East were the second most active in attempts to obtain unauthorized access to sensitive or classified information and technology resident in the U.S. cleared industrial base, as they were in FY12. The number of reported collection efforts linked to the Near East increased over 50 percent in FY13, contributing 18 percent of the total, up from 16 percent in FY12.

The affiliations of collectors cited in industry reporting with a Near East connection remained relatively stable from FY12: all five affiliation categories maintained the same ranking year over year. However, last year's top two affiliations, government-affiliated and commercial, both lost ground to the individual affiliation. Government-affiliated declined from 47 to 44 percent and commercial from 28 to 22 percent, while individual increased from 11 to 17 percent. Nonetheless, all five affiliations increased in number of cases, by margins ranging from 23 to 140 percent.

The shift in reported approaches toward the individual affiliation paralleled the most significant change in FY13 industry reporting related to the Near East: the seeking employment method of operation (MO) joining academic solicitation and attempted acquisition of technology (AAT) among the top three most reported. The number of reports citing seeking employment tripled, and the category more than doubled its share of the total, to nine percent. Added to a near doubling in reported academic solicitation, the top MO, and a resultant increase in its share from 38 to 46 percent of the total, these two methodologies together accounted for over half of FY13 industry reporting. Most cases of academic solicitation continued to consist of Near

East students seeking postgraduate positions, with smaller numbers seeking thesis assistance, reviews of draft scientific publications, and/or access to U.S. research papers.

AAT remained the second most frequently cited MO in FY13. While its share of the total declined from 22 to 16 percent, the number of reported cases increased. Similarly, reported cases of request for information (RFI) increased, but the MO's share declined. However, these two combined still accounted for a quarter of the Near East total.

Among other MOs, foreign visit increased slightly in share, suspicious network activity (SNA) and exploitation of relationships remained steady, while solicitation or marketing services declined. All but the last, however, increased in number of reported cases.

The increased incidence of SNA in Near East-related industry reporting was noteworthy. Not only did the number of reported cases nearly double from FY12, but the last two years of industry reporting to DSS showed that computer network exploitation (CNE) actors from the Near East have increased their level of sophistication. They demonstrated the ability to conduct more technically advanced computer network operations (CNO) against cleared contractor facilities.

Near East academic solicitations had a significant effect on the targeted technology categories, as a large number of students from the region expressed interest in U.S. academic research programs related to computational fluid dynamics (CFD) and associated subdisciplines. This interdisciplinary field spans several sectors of the Industrial Base Technology List (IBTL), but had an especially large role in placing marine, aeronautic, and space systems in the first, third, and fourth positions on the list of Near East-targeted technologies in FY13 industry reporting.

FY12's top technology sector, electronics, interposed itself among the above categories in second place for FY13, with its share declining from 14 to eight percent of the total. Energy systems, however, experienced a profound change, the number of reported cases increasing by a factor of eight, the share rising from one to six percent, and the ranking rising to fifth.

Last year's second-ranking Militarily Critical Technologies List (MCTL) category, information systems (IS) at 13 percent, broke down in FY13 into the IBTL categories of software and command, control, communication, and computers (C4) at six percent apiece. Similarly, last year's third-ranking technology sector at ten percent, lasers, optics, and sensors, broke down in FY13 into separate sectors for each of those technologies plus another for radars, but together they accounted for nine percent of the total.

COLLECTOR AFFILIATIONS

Near East-linked reported collection activity attributed to government-affiliated collectors accounted for 44 percent of FY13 industry reporting, a slight decrease in share from FY12, but still twice as much as the next category, and representing a 43 percent increase in number of reported cases. Government-affiliated collectors continued to be mostly associated with either public universities or government-linked firms. Both patterns involved, in some fashion, seeking and attempting to exploit access to cleared facilities, whether by visiting or attaining some position there.

The number of reported suspicious contacts associated with Near East commercial entities increased during FY13 by nearly a quarter, although this affiliation's share also decreased, from 28 percent in FY12 to 22 percent in FY13. This share for the commercial affiliation was the lowest among the four main collector regions. Near East commercial companies that contacted cleared facilities directly generally sought access to sensitive and/or dual use technology.

However, much of the collection activity traceable to the Near East has historically come through procurement networks consisting of intermediaries and front companies. DSS analysis reaffirmed this activity by linking suspected Near East procurement attempts to commercial entities located in the Near East as well as other regions, including East Asia and the Pacific, Europe and Eurasia, Africa, and North America.

Analyst Comment: Despite the discussion of reported activity above, the Near East's extensive use of procurement networks likely concealed even more such activity, contributing to the commercial affiliation continuing to account for a

relatively low proportion of reporting attributed to the region. Commercial companies from the region that, for a variety of reasons, could not make direct approaches to cleared contractors very likely originated many of their requests from other countries having more favorable trade relationships with the United States. (Confidence Level: High)

Those members of the Near East commercial sector with better access to cleared contractors targeted U.S. information and technology prolifically, based on industry reporting. Defense firms and consultants contacted cleared contractors via email in attempts to acquire U.S. defense technology or met in person with cleared contractor personnel to solicit partnerships or distributorships. As an example, one arrangement proposed in January 2013 would have granted Near East access to radar products that are restricted under the International Traffic in Arms Regulations (i.e., ITAR). The cleared contractor rejected the proposal because the Near East commercial entity in question did not meet U.S. standards for export control of defense items.

Analyst Comment: Accepting a Near East company's solicitation to serve as an overseas distributor very likely can appear to a cleared contractor to be a legitimate and economically sound way to expand its customer base. However, depending on a foreign commercial entity that is unable to secure controlled items and unaware

of export regulations would almost certainly risk unauthorized access to sensitive or classified technology. (Confidence Level: High)

Industry reporting reflects that some of the Near East commercial entities that most persistently target U S. information and technology cooperate with national intelligence services, and thus provide a conduit for collection attempts against U.S. sensitive, classified, and export-controlled technology. Some of these firms allow intelligence services to include intelligence officers (IOs) within company delegations that visit cleared facilities, and even accept taskings to target specific cleared contractors. Embedded IOs during FY13 included defense, military, and other government personnel.

Analyst Comment: Foreign visits associated with foreign military sales or agreements create a legitimate need to incorporate military or other government personnel into visiting delegations. Near East intelligence services very likely attempt to exploit the resultant access by including IOs to collect information and assess cleared contractor personnel. (Confidence Level: Moderate)

Individual collectors accounted for 17 percent of suspicious contacts in FY13 industry reports, an increase from 11 percent the previous year. This reflected an increase in the number of cases of over 140 percent. These individuals contributed heavily to the prominence of the academic solicitation and seeking employment MOs. In many of these

Figure 10: Collector Affiliations

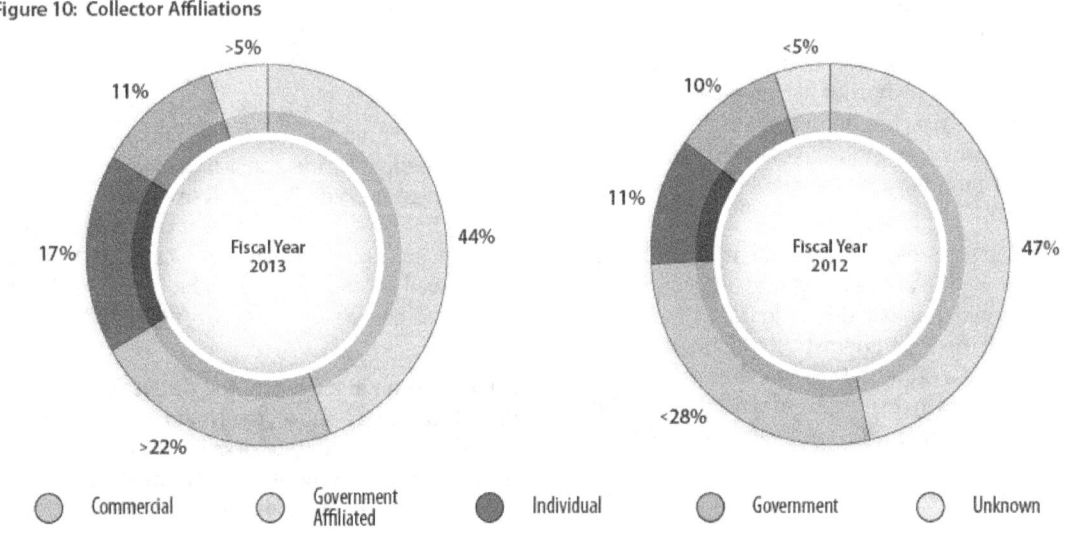

cases, Near East individuals making suspicious contacts to cleared facilities omitted any affiliation with a particular government agency, government-affiliated university or research facility, or commercial entity.

Analyst Comment: Senders probably omitted additional contact information associated with their country of origin or affiliations in an attempt to increase the likelihood the cleared contractor would engage in further discussion. (Confidence Level: Moderate)

METHODS OF OPERATION

During FY13, academic solicitation remained Near East entities' most commonly reported MO, accounting for 46 percent of industry submissions. The number of reported cases increased more than 80 percent over the previous fiscal year. Near East academic solicitation continued to account for a significantly higher share of the total than other MOs—the next nearest category was AAT at 16 percent. The seeking employment MO, which shares similarities with academic solicitation, tripled in number of reported cases from FY12, increased its share of the total from four to nine percent, and became the third most reported MO.

It is noteworthy that reported Near East attempts to attend U.S. universities increased despite pervasive and continuing economic difficulties throughout the region. Resultant strictures in funding support would tend to threaten financial hardships for Near East students studying in the United States. Nonetheless, significant numbers of Near East students continued to study in the United States during the 2012-13 academic year.

The majority of academic solicitations to cleared industry involved Near East students seeking postgraduate research positions at U.S. universities involved in sensitive and/or classified research for the Department of Defense (DoD). The remaining solicitations typically involved requests for thesis assistance, review of draft scientific publications, and/or access to U.S. research papers. According to IC reporting, some Near East regimes leverage academics (professors, students, and researchers) to exploit academic practices and generous U.S. student visa laws in order to facilitate collection efforts against emerging U.S. DoD and civilian technical research.

The following example describes a Near East student's likely attempt to exploit the U.S. academic community to benefit his country. In September 2012, a national of a Near East country contacted a cleared contractor seeking a position relating to critical infrastructure security. He requested to see the criteria the professor used to select students.

The individual's curriculum vitae and IC reporting identified him as having held positions and been an active participant in his country's computer testing, network security, and hacking communities. Open-source research disclosed that he was scheduled to speak in the United States in January 2013 at a symposium on related subjects that would be attended by government employees and cleared contractors.

According to IC reporting, while the individual's country does not currently possess the capability to attack the type of critical systems on which he appears to concentrate, its computer scientists and other cyber actors have persistently conducted research on the vulnerabilities of and possible exploitation methods applying to such systems, which extraction and production industries use worldwide.

Analyst Comment: A body of collateral intelligence reporting revealed the individual in question to be well known throughout his country's computer network security and hacking communities. If he were to attend a U.S. university, he would likely gain opportunities to study security practices and identify vulnerabilities of U.S. networks. Any such knowledge gained would likely enhance Near East entities' capabilities to attack critical U.S. systems. (Confidence Level: High)

Near East IOs probably use information they obtain about students from applications to study abroad as well as participation in relevant student and university activities to spot and assess those studying in programs or developing expertise of interest to Near East governments. (Confidence Level: Moderate)

AAT was the second most reported Near East MO in FY13. While the volume of suspicious contacts remained similar to FY12 numbers, the category's percentage decreased from 22 to 16 percent. Some procurement techniques practiced in the region, such as the use of witting and unwitting intermediaries located in various countries and

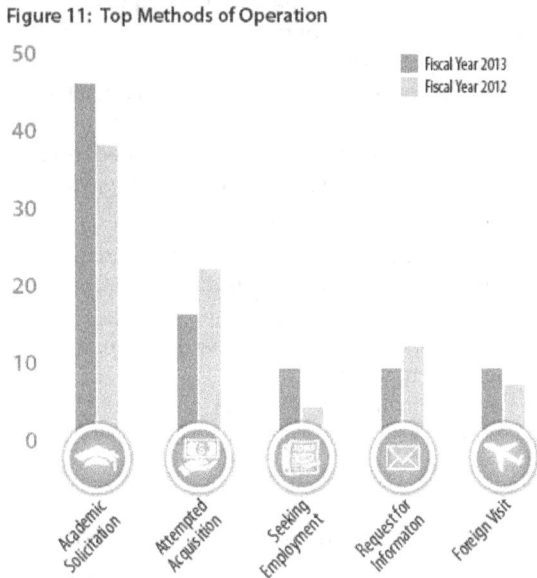

Figure 11: Top Methods of Operation

Fiscal Year 2013
Fiscal Year 2012

Academic Solicitation / Attempted Acquisition / Seeking Employment / Request for Information / Foreign Visit

falsified end-user statements, make it increasingly difficult to attribute requests to purchase U.S. technology to Near East entities.

Nonetheless, industry reporting showed that in FY13, similar to last year, Near East entities attempted to acquire a variety of export-controlled electronics and to download U.S. export-controlled software from cleared contractor websites. An additional body of IC reporting indicates Near East procurement agents are currently very active in attempts to acquire U.S. military and dual-use components.

Similarly, the number of industry reports citing the RFI MO remained relatively consistent in number from FY12, but the MO's share of the total declined from 12 to nine percent in FY13, and it fell from third to fourth in ranking. Near East entities typically sent emails seeking information regarding export-controlled technology. Requestors were often reluctant to provide end-user or end-use data, obscuring the ultimate recipient.

Near East entities with access to cleared contractor facilities attempted to leverage foreign visits to access U.S. information and technology during FY13. Industry reports of attempts to exploit established or emerging defense agreements via official delegation visits to cleared facilities demonstrated substantial growth over FY12, increasing 80 percent in number and from seven to nine percent of FY13

reported incidents. In contrast, contacts that DSS characterized under the solicitation or marketing services MO, usually consisting of offers to serve as an overseas distributor for a U.S. defense platform, declined in number by nearly a quarter, and the MO's share was halved. Thus, in a context of reduced budgets, analysis of industry reporting showed Near East entities spending fewer resources on fostering new business relationships but more effort on exploiting existing relationships.

When possible in FY13, Near East entities leveraged personal contact when targeting U.S. information and technology. Such approaches included intelligence and military entities attempting to exploit a personal exchange or encounter. Venues included cleared contractor facilities and defense conferences. Techniques included attempting to conceal IOs' status, aggressively soliciting for restricted technical data, and attempting unauthorized use of cameras, cell phones, and laptops.

Analyst Comment: DSS assesses that the increase from FY12 in numbers of reported incidents involving personal contact was very likely not due to an increase in Near East collection activity but instead to improved cleared contractor awareness of collection techniques typically employed during foreign visits and U.S. and international defense conferences. Near East actors probably used public venues and personal exchanges as exploitable opportunities to attempt to solicit and elicit sensitive or classified U.S. information and technology. (Confidence Level: Moderate)

An example occurred during a February 2013 visit of a delegation combining representatives from two government-affiliated Near East firms to a cleared facility. During the visit, observers saw an individual, later identified as a suspected IO, with his cell phone and an alleged recording device in the facility.

Analyst Comment: Collectors very likely seek opportunities to manipulate or circumvent normal security procedures to facilitate IO access to U.S. information and technology. (Confidence Level: Moderate)

Near East countries that receive U.S. military technology—whether licitly or illicitly, and whether those acquisitions are subject to limiting agreements or not—have histories of further transferring military technology to countries of concern outside the region. The pattern includes Near East commercial efforts to acquire U.S. technology specifically on behalf of third entities, a trend that continued in FY13 AAT statistics.

In a case from March 2013, a Near East electronics distributor requested to purchase multiple variants of a cleared contractor's military-grade accelerometers for an unidentified end user. With regard to variants and quantities, these requests were consistent with numerous others originating from a third country of concern. Significantly, the Near East company in question had previously procured military-grade technology for an extraregional company by falsely claiming the end user was another firm from its own country.

In FY13, industry reports of Near East SNA almost doubled in number of cases. It is even more worrisome that, while in FY11 SNA originating from the region was still limited to less technically sophisticated spear-phishing and social networking site activity, in FY13 Near East cyber actors expanded and upgraded their activities to include more advanced attacks. This coincided with an uptick in the level of success for Near East SNA, including in the seriousness of the intrusion and the number of exfiltrations of data from cleared contractor networks.

IC reporting indicates that Near East CNE programs continue to evolve. Their tactics, techniques, and procedures (TTPs) are improving, and they use external resources to enhance their operations. SNA actors within particular countries share tradecraft in research, programming, and network exploitation. Some of them are associated with their governments, others with information technology companies.

Analyst Comment: Based on their recent employment of vectors of increasing technical complexity, Near East cyber actors have very likely capitalized on experience garnered from previous SNA attempts. This experience curve, which governments in the region have supported with concerted efforts, has probably led to an increase in the level of sophistication of CNE programs and an increase in cyber actors' SNA success rate. (Confidence Level: Moderate)

TARGETED TECHNOLOGIES

In FY13, Near East collection efforts spanned all categories of the IBTL, just as they spanned all categories of the MCTL in previous years. The region's collectors most commonly targeted information and technology related to marine systems, electronics, and aeronautic and space systems. Together these four categories accounted for a third of the total.

In addition to the shuffling of technology categories and reporting data occasioned by DSS' shift from the MCTL to the IBTL, some additional sifting of data can be laid to improved attribution of collection efforts. For instance, industry reports on the general category of electronics declined in number while those citing the specific categories of marine, aeronautic, and space systems all increased.

During FY13, industry reporting on Near East academic solicitations heavily influenced the apportionment between IBTL categories: the most reported technology sectors tended to be linked to student interest in gaining entry to specific U.S. research programs. A large component of Near East student applicants requested to conduct research related to CFD.

Fluid dynamics is a subdiscipline of fluid mechanics that deals with fluids in motion and can be subdivided into aerodynamics and hydrodynamics. Due to the interdisciplinary nature of CFD, DSS divided foreign expressions of interest in the academic programs into the primary research applications of marine, space, and energy systems. This caused Near East-connected targeting reports citing all three of these IBTL sectors to fall within the top five in number of industry submissions in FY13.

The following example details a Near East student's requests to various U.S. universities to conduct research regarding CFD, propulsion, and energetic materials. During the fiscal year, he sent many academic solicitations to the same professors who are currently receiving high interest from his compatriots. From October 2012 through September 2013, he applied to four separate U.S. universities seeking a research position. In his emails to professors in cleared programs, he stated he was interested in CFD, fluid mechanics, and similar subjects. The professors are subject matter experts in various fields related to propulsion.

Figure 12: Top Targeted Technologies

Marine Systems 10%
Academic programs
(computational fluid
dynamics)

Electronics 8%
Radiation-hardened
integrated circuits;
monolithic microwave
integrated circuits

**Aeronautic
Systems 8%**
Manned, fixed-wing
aircraft; unspecified
source code information;
unmanned aerial system
platforms

Space Systems 7%
Academic programs
(computational fluid
dynamics)

Near East entities continued to display interest in electronics, although relevant FY13 industry reporting showed a decline of more than 20 percent in number of cases from FY12, and from 14 to eight percent in share of the total, dropping the MO to second place. FY13 industry reporting portrayed Near East entities as focusing on a variety of dual-use electronic components. These components included types of integrated circuits whose manufacture the IC has assessed to be beyond the Near East's industrial capacity. These components became the subject of collection efforts involving academia, front companies and third parties, and CNO.

Aeronautic systems technology was Near East collectors' third most targeted category in FY13, based on reporting data that was comparable to FY12's. Whereas in previous years, unmanned aerial systems (UAS) were the most commonly targeted aeronautic platforms, FY13 industry reporting indicated a shift toward attempts against manned, fixed-wing aircraft of multiple types that would be useful in any warfare against likely regional opponents. Documented Near East collection attempts primarily sought unspecified source code information.

Previous cleared industry reporting had not supported findings reported elsewhere in the IC concerning focused targeting of these aircraft. However, FY13 industry reports showed greater correspondence. For example, in February 2013, Near East personnel with access to a cleared contractor facility producing fixed-wing aircraft attempted to violate security policies by moving throughout the facility without their required escort. Other Near East

solicitors subjected DoD personnel supporting fixed-wing aircraft production to requests for restricted data associated with an IS platform to which the solicitors had not been granted access.

Analyst Comment: Cleared industry reporting previously lacked documentation of Near East targeting of technology specific to production of the aircraft in question. This dearth was inconsistent with IC reporting that the aircraft constituted a collection priority. The FY13 increase in industry reports of attempts directed against the aircraft was almost certainly more indicative of improved cleared contractor awareness than a sudden reallocation of Near East collection efforts against this platform. (Confidence Level: High)

Notwithstanding the demonstrated interest in manned aircraft, Near East defense firms continued collection efforts against UAS platforms as well. Unauthorized photography and persistent solicitations for sensitive technical data on tactical reconnaissance UAS platforms occurred at defense conventions. In FY13, cleared industry reporting provided evidence that collection entities tasked Near East companies with collecting specific information on identified U.S. companies and technologies at named defense conferences.

Analyst Comment: Given the access to various U.S.-developed UAS platforms that defense conferences offered, they almost certainly provided Near East entities the most opportunity-rich environment available for their collection efforts. (Confidence Level: High)

Near East entities continued to express interest in various U.S. software programs during FY13. Targeted programs related to a wide range of applications; however, similar to last year, the majority of software requests consisted of Near East entities seeking to download from a cleared contractor's website a particular software suite used for space, defense, and intelligence systems. In their attempts to gain access to this software, Near East entities often falsely claimed U.S. cities and organizations in their end-user information.

Analyst Comment: The Near East collection entities in question almost certainly provided the falsified end-user information in an attempt to obfuscate their countries of origin. (Confidence Level: High)

In FY13, Near East entities focused considerable collection efforts against C4 technology, primarily airborne and vehicle-based radio platforms. In previous years, DSS categorization methodologies assigned such attempts to the larger IS category, which in FY12 accounted for 13 percent of the Near East-linked total. In FY13, C4 alone registered over half as many cases as all of IS in FY12, and accounted for six percent of the total. Most reported attempts sought to leverage access to cleared facilities and defense conferences.

The subject of this publication's special focus area, inertial navigation systems (INS), falls within the positioning, navigation, and time (PNT) technology category. PNT accounted for only one percent of FY13 Near East-linked reporting, and Near East entities accounted for only 18 percent of cleared industry submissions concerning collection activity targeting U.S. INS technology. However, IC reporting indicates a persistent Near East interest in PNT technology. Near East collectors' use of front companies may have obscured their link to some of the remaining reported targeting.

OUTLOOK

The hierarchy of affiliations in industry reporting attributed to Near East collectors will likely remain unchanged during the next fiscal year. A high percentage of reported Near East-connected cases will likely continue to involve government-affiliated collectors targeting cleared industry. Some regional collection efforts will likely continue to employ procurement agents, brokers, and front companies in their attempts to acquire U.S. technology. (Confidence Level: Moderate)

Near East-based students and government-affiliated researchers will likely continue to attempt to exploit the academic community by soliciting assistance on fundamental and developing research and attempting to acquire dual-use components under the guise of academic research. (Confidence Level: Moderate)

Near East military, other government, and defense industry components will almost certainly continue to work together to target U.S. technology. As portrayed in previous cleared industry reporting, this collaborative effort will very likely continue to attempt to leverage established defense agreements and existing partnerships. To the extent that Near East entities enjoy access to cleared contractor facilities, they will very likely continue to attempt to leverage and exploit it. Visits to cleared facilities will almost certainly continue to provide opportunities for Near East military and other government components to attach intelligence personnel to official delegations. Any success at embedding IOs within those delegations will very likely create opportunities for unauthorized access to sensitive or classified U.S. information and technology. (Confidence Level: High)

Near East consultants will probably continue to solicit partnerships with U.S. cleared contractors and attempt to procure export-controlled technology for their countries' defense firms and militaries. Conversely, as Near East relations with third countries of concern outside the region continue to develop, the resultant increased interactions will likely be reflected in cleared industry reports of Near East consultants acting as a procurement conduit for defense items prohibited to other countries. (Confidence Level: Moderate)

Given the increased success of Near East SNA at compromising networks and gaining illicit entry into cleared contractor systems, the region's CNE actors in FY14 will likely increase their targeting of cleared industry, using a wide spectrum of vectors. Now that these cyber actors have emerged as a threat in cyberspace, they will likely continue to exercise their newly acquired skills and seek to expand into more threatening CNE activities. (Confidence Level: Moderate)

The technologies Near East entities target will likely depend on their countries' economic situations. Economic strictures, defense budget reductions, and military research and development retrenchments may reduce or eliminate funding for at least some defense-related programs. In such a fluid situation, Near East targeting of U.S. technology will probably remain varied. (Confidence Level: Moderate)

Your Account has been Referred to a Collection Agency

The following case study exemplifies the TTPs and detailed efforts Near East collection entities employ in their attempts to obtain unauthorized access to sensitive or classified information and technology resident in the U.S. cleared industrial base, in this case unmanned aerial vehicles (UAVs).

During a 2013 UAV conference, an unidentified individual dropped a three-page document on the convention floor. The document was in a Near East language. Translated, the title referred to "Information Collection during the…Conference and Exhibition."

Contextual clues within the document provided strong evidence it was prepared by or for a particular Near East company. The document tasked three nationals of the same country as the Near East company to collect specific information on prioritized U.S. companies and technologies at the conference. One of the individuals so tasked was previously unknown to DSS and had not appeared in IC reporting. However, the document contained contextual clues that he was an employee of the Near East company.

The document instructed collectors on which companies to target, what information to collect and report, and how to do so. The collectors were to attend relevant events; take photographs; obtain copies of presentations; clarify terms; collect product lists, brochures, and advertising material; and seek client information.

Another individual tasked in the document to acquire information on defense contractors was associated with a different Near East company that is also a component of the larger company. This company supplies data and video links for UAV systems and other advanced military wireless communications products. The company is involved in the design, development, and marketing of systems similar to the UAV technologies on which the first individual was directed to focus.

Analyst Comment: DSS assesses that the Near East company probably prepared the document as an aid for its employees attending the UAV conference. DSS further assesses that the U.S. companies and specific technologies targeted in the document approximately correspond with business sectors with which the Near East company is involved. The tasked individuals very likely sought to collect information on cleared contractor business development and technological advances during the conference to aid their company's product-development and marketing efforts. (Confidence Level: Moderate)

SOUTH & CENTRAL ASIA

OVERVIEW

Fiscal year 2013 (FY13) cleared industry reporting showed that South and Central Asia collectors continued to demonstrate interest in a wide variety of U.S. information and technology. The number of reported incidents possessing a South and Central Asia nexus increased by nearly 60 percent over FY12, accounting for 16 percent of all reports from cleared industry in FY13, up from 12 percent the previous year. South and Central Asia remained the fourth most attributed region in FY13 reporting of foreign collection attempts to obtain unauthorized access to sensitive or classified information and technology resident in the U.S. cleared industrial base. However, the percentage increase in the number of cases attributed to this region from FY12 to FY13 was the highest rate of the four main collector regions.

World economic conditions and resultant defense budget retrenchments hampered the ability of South and Central Asia regimes to pursue desired military modernization via both direct purchase and independent indigenous research and development (R&D) efforts. Therefore, targeting U S. information and technology to contribute to the development of indigenous military production capabilities and reverse-engineering foreign defense systems continued to present an attractive path toward improving their militaries' technical capabilities.

After earlier bumps in the road, U.S. relations with South and Central Asia generally improved over the course of FY13. However, in several cases relations between South and Central Asia countries and problematical third countries outside the region also improved. From strategic partnerships to economic development pacts to technology development and acquisition assistance agreements to general expressions of cooperation, these third-country relations represented continuing risks of the transfer of U.S. technology. Some South and Central Asia countries have provided U.S. technology to third countries in the past, and may feel obligated to do so

in the future as quid pro quos for the benefits they gain as the relationships develop. Thus, these ties between South and Central Asia and third countries in other regions represent a significant threat to any U S. defense technology their collectors acquire, whether legally or illicitly.

South and Central Asia's long-standing internal domestic and intra- and extra-regional frictions have made it a volatile region. Overlaying the evolving international relationships described above on top of these existing frictions is likely to make it even more volatile. Some national militaries may attempt to use these existing frictions and potential conflicts, perhaps taking advantage of any seemingly provocative actions, to leverage support for aggressive collection activity against sensitive or classified information or technology resident in the U S. cleared industrial base.

The most noticeable change in FY13 industry reporting related to South and Central Asia was an increase in attempts to gain employment, internships, and research positions at cleared facilities, a trend the Defense Security Service (DSS) anticipated in last year's version of this publication. In FY12, attempted acquisition of technology (AAT) was the most reported method of operation (MO) at 31 percent of the total; however, the substantial surge in FY13 industry reports citing the seeking employment and academic solicitation MOs dropped AAT to third, accounting for 16 percent of the total. The combined number of reports involving résumé submissions nearly doubled. Together these two MOs accounted for nearly 70 percent of the total.

Along with AAT's decrease in share from 31 to 16 percent, request for information (RFI), the fourth most cited MO, decreased from 18 to eight percent, and solicitation or marketing services followed at six percent, down from eight percent in FY12. As in FY12, these cases mostly involved commercial entities serving as procurement agents by attempting to obtain technology for South and Central Asia militaries or other government organizations, sometimes in response to government tenders.

While in FY12 government-affiliated entities accounted for the largest portion of industry reporting with a South and Central Asia nexus, the number of cases associated with individuals more than quadrupled in FY13. This notable surge reflected the significant increase during FY13 in the number of résumés submitted to cleared facilities for positions working under subject-matter experts (SMEs). Individuals accounted for 36 percent of the FY13 total, higher than both government-affiliated and commercial entities, which accounted for 33 and 24 percent, respectively. With defense budgets decreasing in many cases, South and Central Asia efforts may have shifted away from independent indigenous development or direct purchase from cleared contractors toward attempting to gain access to U.S. military technologies while they are still under development.

In FY13 South and Central Asia-connected contact reports, electronics remained the most requested technology at nine percent of the total, although this was down from 15 percent in FY12. Beyond that, DSS' change in technology categorization methodology from the Militarily Critical Technologies List (MCTL) to the Industrial Base Technology List (IBTL) slightly shifted the next technology sectors represented in the listing. Whereas lasers, optics, and sensors (LOS) was the second most cited MCTL category in FY12 at 14 percent, in FY13's IBTL listing radars (formerly part of LOS) alone accounted for four percent of South and Central Asia-connected reporting (optics and sensors (acoustic) together contributed an additional two percent). Command, control, communication, and computers (C4) was formerly part of the information systems (IS) category. In FY12, IS was the third most cited category at 13 percent; in FY13, C4 on its own was the second most requested technology category, at five percent of industry reporting.

Among the next most cited technologies, nanotechnology, aeronautic systems, and marine systems together accounted for an additional ten percent of the total. Another nine sectors registered in the single digits, but 39 percent of reports fell into the unknown category.

Since the majority of incidents attributed to the individual affiliation involved South and Central Asia nationals submitting résumés to various cleared contractors without specifying precise specializations or positions, it remained unclear which specific U.S. technologies they were targeting, if any. When employment solicitations did specify a U S. technology, they focused on nanotechnology, materials (raw and processed), and marine systems. Consistent with FY12, non-résumé solicitations in FY13 industry reporting focused on electronics technology—the top target overall—as well as C4 and radars.

COLLECTOR AFFILIATIONS

In FY13 industry reporting, 36 percent of South and Central Asia-connected incidents involved individuals not linked to commercial or government-affiliated entities. The number of reports attributed to individuals increased by more than a factor of three from FY12.

Defense budget reductions in South and Central Asia presented a number of challenges to regional defense industries. One of the longer-lasting effects is likely to be cutbacks to R&D budgets, including the ability to recruit and hire graduates in high-tech fields to conduct R&D in military technologies over the long term. Particular areas likely affected include missile, aerospace, electronic warfare (EW), and infantry support programs.

Analyst Comment: There is an even chance these budget cuts contributed to the surge in requests from South and Central Asia nationals for research positions, internships, and positions within cleared contractor facilities. While it is unlikely that individual applications constituted responses to government targeting initiatives, any information or technology shared with the individuals could fulfill government gaps when they return home. (Confidence Level: Moderate)

The individuals represented in industry reports were primarily job seekers who applied for positions throughout cleared industry, often in response to vacancy announcements from the cleared contractors that clearly stated requirements for U S. citizenship or a security clearance for sensitive positions. While these individuals were therefore ineligible for the positions, the significant and increased volume of South and Central Asians attempting to gain employment at research centers and cleared facilities was nonetheless noteworthy.

Analyst Comment: While most of the individuals making these contacts were probably legitimately interested in obtaining positions at cleared contractor facilities, such placements offer opportunities to exploit access to personnel, information, and technologies. (Confidence Level: Moderate)

In a typical case, between January and March 2013, a South and Central Asia national sent numerous emails to the same cleared contractor requesting a graduate research position at the cleared facility.

Analyst Comment: While observers reported no illicit actions by the applicant in question, given the aggressive and persistent nature of his solicitations to research under a cleared SME, there is an even chance he sought access to sensitive or classified research on behalf of unauthorized foreign entities. (Confidence Level: Moderate)

Industry reports attributed to government-affiliated entities increased nearly 50 percent in FY13. As in FY12, the actors primarily consisted of students, researchers, and professors from regional institutions of higher learning. Cooperation between such institutions and South and Central Asia governments aims at augmenting available indigenous scientific talent; however, cumulative reporting shows that

governments also work through such institutions with the intent of obscuring their footprint while attempting to acquire sensitive technologies from abroad, including the United States.

Analyst Comment: South and Central Asia students and technical graduates persistently pursue postdoctoral, research, and internship opportunities under technical SMEs employed in cleared industry in the United States. There is an even chance such students intend thereby to gain access to sensitive or classified information or technology for further transfer to their home countries. (Confidence Level: Moderate)

Analysis of cleared industry reporting connected to South and Central Asia revealed that another segment of reports attributed to government-affiliated entities consisted of commercial companies attempting to procure technology on behalf of national military services or other government organizations. As part of government procurement processes, South and Central Asia military and other government entities have historically posted tenders online, allowing both government-affiliated entities and commercial companies to bid. Similar to FY12, DSS corroborated that a number of government-affiliated entities were seeking U.S. technologies that would have fulfilled tenders consistent with particular armed services' current modernization efforts.

Figure 13: Collector Affiliations

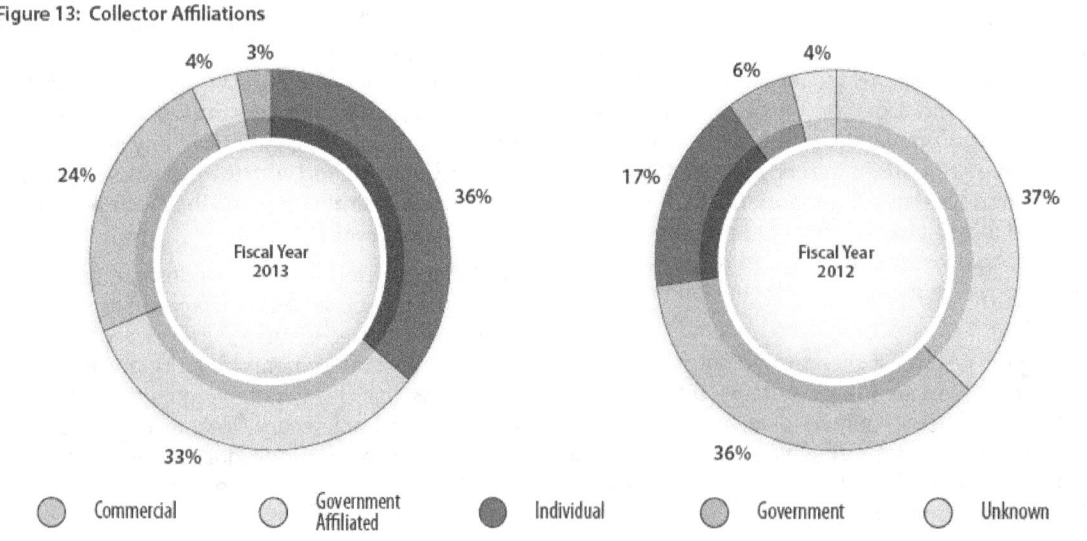

Analyst Comment: The decrease in South and Central Asia defense budgets likely provides part of the explanation for the decrease in the share of reports attributed to government-affiliated entities. Regional military services and other government organizations likely advertised fewer tenders this fiscal year due to financial constraints. (Confidence Level: Moderate)

Cleared industry received some solicitations from defense firms owned by South and Central Asia governments. And, as noted above, the most prominent pattern involving government-affiliated entities consisted of commercial companies attempting to procure U.S. technology in response to official government tenders. Yet the weight of available evidence portrays technology procurement processes in the region as complex and protracted, to the point that they hinder the procurement efforts even of high-profile defense programs intended to provide high-priority military capabilities.

Therefore, another less prominent yet still noteworthy pattern documented in industry reporting involved entities in third countries, including the United States, responding to these publicly available government tenders. South and Central Asia collection efforts use such foreign countries as well as front companies, domestic middlemen, and non-intelligence end-user companies in their targeting of U.S. defense technology. These efforts employ the RFI and AAT MOs with falsification of information to mask true end-user identities.

Analyst Comment: Willingness to use nontraditional collectors almost certainly facilitates South and Central Asia entities' employment of collection techniques such as leveraging nationals living abroad to send U.S. technology home while falsifying end-user certifications. This approach almost certainly provides the region's governments with some degree of disassociation from illicit procurement activities. (Confidence Level: Moderate)

Some national defense officials within South and Central Asia openly admit that their defense establishments, including associated research facilities, practice reverse-engineering of acquired foreign technologies, and that they intend to continue to pursue such efforts even as their own indigenous capabilities improve. A possible example occurred in December 2012, when a representative of a South and Central Asia company contacted a cleared contractor requesting to purchase a single export-controlled laser device. Research identified a publicly available defense tender that matched the requested specifications and listed the end user as a government agency.

Analyst Comment: Acquisition of U.S. technologies via the publicly available defense tender process is an overt method of procurement. But in this case, since the requestor sought only one item and a government agency may have been the ultimate recipient, the item was very likely destined for reverse-engineering in pursuit of enhanced indigenous development capabilities. (Confidence Level: Moderate)

Commercial entities accounted for 24 percent of all South and Central Asia-connected industry reporting in FY13. This percentage represented a decrease from 36 percent in FY12, due to the larger increases in the number of reports attributing collection attempts to individual and government-affiliated entities. Nonetheless, the number of reports attributed to commercial entities increased ten percent. Contacts from commercial entities mainly consisted of RFIs and AATs directed toward cleared contractors via email messages and web-card submissions, and approaches during conferences, conventions, and trade shows.

Consistent with FY12 industry reporting, a number of commercial entities attempting to procure technology identified South and Central Asia military or other government organizations as the end users. Commercial companies often bid on tenders these entities posted online. In a significant number of such incidents, DSS observed a pattern in which several South and Central Asia commercial companies solicited the same cleared contractor with identical technology requests, all failing to identify an end user or use. Often, DSS research yielded no publicly available government tender corresponding to these requests. In some instances, DSS identified the requesting commercial companies as procurement agents either for a governmental organization on a U.S. government restricted end-user list or an organization involved in weapons development.

For instance, in March and April 2013, two South and Central Asia companies contacted the same cleared contractor requesting the same quantity of electronic components, with neither providing an end user. Both companies have long histories of contacting cleared industry, and both are procurement agents that supply items in support of weapons of mass destruction (WMD) and nuclear programs, including to organizations on the U.S. Department of Commerce's Entity List.

Analyst Comment: Given the lack of a corresponding tender and the similarities between different requests from multiple commercial entities, DSS assesses that these entities were likely procurement agents or front companies attempting to procure technology on behalf of South and Central Asia organizations on the Entity List or organizations involved in weapons development. (Confidence Level: Moderate)

METHODS OF OPERATION

FY13 saw a surge over FY12 in reports of seeking employment and academic solicitations, increasing in number of cases by 490 and 55 percent, respectively. These top two MOs accounted for over 65 percent of all industry reports of incidents attributed to South and Central Asia entities. In FY12, these techniques accounted for 37 percent combined, so the FY13 results demonstrated a significant increase from an already substantial percentage. AAT and RFI remained commonly used MOs, accounting for 16 and eight percent of the total, respectively.

The reported incidents of seeking employment and academic solicitation largely consisted of South and Central Asians requesting research positions at cleared contractor components of academic institutions and applying for positions in cleared industry that often required U.S. citizenship or a security clearance. While the significant majority of these approaches appeared to be overt attempts to actually obtain positions within U.S. cleared industry, industry reports documented some South and Central Asia individuals aggressively seeking information regarding certain positions.

In one example from September 2013, an identified national of a South and Central Asia state contacted a cleared contractor employee requesting detailed information, including personnel identifications, associated with the cleared facility's information technology (IT) security team. The individual claimed he had found the cleared employee's contact information on a business networking profile that listed the employee as the IT security manager. After ending the conversation, the cleared contractor employee—who was not, in fact, the IT security manager—confirmed that his profile neither cited such a position nor divulged his current place of employment. Furthermore, according to the cleared employee's networking profile record, the South and Central Asian had never viewed his account as claimed.

Analyst Comment: Although the inquiring individual's information was erroneous, he very likely used it to provide an opening to obtain details on the cleared position. Given that the cleared employee's employment information was not accessible via the means claimed and the business networking account did not reflect the South and Central Asia national having viewed the cleared employee's profile, the solicitation was probably a targeting attempt. (Confidence Level: Moderate)

Reports citing the seeking employment MO demonstrated no discernible pattern beyond the many cases of individuals responding to publicly available vacancy announcements. South and Central Asians solicited a wide variety of cleared contractors seeking employment opportunities, including companies specializing in telecommunications services, network systems development, combat vehicles development, and optics technologies. A majority of the prospective employees were mechanical, systems, or electrical engineers.

Analyst Comment: These increased attempts by individuals from the region to obtain employment may have been legitimate; however, employment at a cleared contractor could provide individuals access to sensitive or classified U.S. information and technology. Given the reduction in defense budgets within the South and Central Asia region, these employment solicitations could provide the countries involved with offsetting means to obtain U.S. technology. Any knowledge South and Central Asians gain while employed at cleared contractors

likely contributes to the expertise required to bolster indigenous production capabilities. (Confidence Level: Moderate)

AATs and RFIs are very similar in use and direct in nature. AATs and RFIs both decreased in number from FY12 to FY13. Collectively they accounted for about a quarter of South and Central Asia-connected industry reporting in FY13, a considerable decrease from accounting for nearly half in FY12. This was largely due to the greater surge in employment solicitations.

AAT/RFI incidents essentially consisted of South and Central Asia companies (commercial or government-owned) emailing cleared contractors to request sensitive information, such as pricing or technical specifications, or to attempt to acquire export-controlled technology, usually specific components or platforms. Some South and Central Asia companies that had previously been prominent in attempts to acquire U.S. technology became less willing to provide end-user data in FY13, and DSS research could not validate applicable government tenders, which are usually accessible.

Analyst Comment: Given DSS inability to corroborate these acquisition efforts via tenders available to government procurement agents, the agents likely deliberately sought to obscure the end use. (Confidence Level: Moderate)

Industry reporting reflected South and Central Asia government organizations' use of commercial entities as procurement agents employing RFIs. In March 2013, a representative of a South and Central Asia organization that conducts engineering research contacted a cleared contractor requesting a variety of product information on the contractor's radar, EW, and communications products. Past reporting documents a history for the organization as a procurement agent for a government entity that is responsible for developing and producing nuclear, conventional, and unmanned weapons and delivery systems. The organization has used third parties to acquire technology in the past. Three days later, the same cleared contractor received a request for various electronic components from a different procurement agent for the same organization.

Analyst Comment: Based on the second procurement agent's history, he could have intended the requested items for a number of different military or other government end users. However, based on the short interval between the requests as well as corroborative reporting, DSS assesses there is an even chance the second agent requested the technology for the same end user as the first, likely the engineering research organization or a similar entity. These entities probably used a second procurement agent in an attempt to obscure the actual end user. South and Central Asia procurement agents very likely view AATs and RFIs as productive means of obtaining U.S. information and technology. (Confidence Level: Moderate)

South and Central Asia companies in FY13 also employed solicitation or marketing services, a category that increased in number of reports but decreased slightly in its share of the total to six percent. These incidents primarily consisted of solicitations to cleared industry to market a cleared contractor's technology to regional customers. A number of these commercial companies work with regional ministries of defense and armed forces. While these requests often appeared legitimate, many of the companies involved had direct affiliations with entities on the Department of Commerce Entity List.

Analyst Comment: While these business solicitations may have been legitimate, any resulting relationships could provide South and

Figure 14: Top Methods of Operation

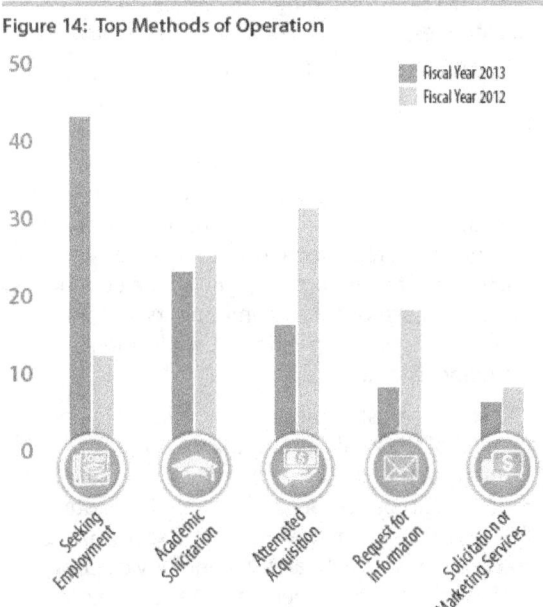

Central Asia entities access to sensitive, classified, or proprietary information and technology resident at the cleared contractor. Any acquisition of such information likely enables the entities to bolster indigenous technology production capabilities. (Confidence Level: Moderate)

Defense partnerships between the United States and South and Central Asia are on the advance. Cleared industry reporting often reflects attempts to leverage these defense partnerships, most often via the exploitation of relationship and foreign visit MOs. While the overall percentage attributed to these MOs was consistent with FY12 industry reporting, the combined number of reported incidents increased by 44 percent. In FY13, such incidents primarily consisted of commercial and governmental entities attempting to leverage foreign visits to cleared facilities to gain unauthorized access to U.S. information and technology, as anticipated in last year's version of this publication.

For instance, in June 2013, a representative of a South and Central Asia company attempted to schedule a meeting during a European air show with cleared contractor executives. Industry reporting has documented various attempts by the company in question to obtain sensitive or classified information and technology from cleared contractors. Further, Intelligence Community (IC) reporting has linked the company to the procurement of export-controlled U.S. technology for restricted end users, including space and WMD programs.

Analyst Comment: DSS assesses that any information the South and Central Asia company obtained during the foreign visit would very likely have been used to advance its country's WMD program or develop capabilities to meet additional military- or other government-determined requirements. (Confidence Level: High)

TARGETED TECHNOLOGIES

General intelligence collection portrays that, while South and Central Asia governments desire to enhance their countries' indigenous technological capabilities, their armed forces continue to prepare for potential conflicts with both intra- and extra-regional opponents. Overall, they seek to gain a qualitative military technology edge over rivals, boost their existing operational defenses, and bolster their defense industries.

Analyst Comment: Developing a defense industry requires significant technological know-how. South and Central Asia collectors very likely view their prominent use of academic solicitation and seeking employment as means of gaining this expertise—whether legitimately or illicitly. (Confidence Level: Moderate)

Acquisition of foreign military platforms can fill near-term gaps; subsequent exploitation of these technologies through reverse-engineering can enhance future R&D efforts. Many incidents of South and Central Asia entities requesting specific U.S. systems involved small quantities of technologies. Analysis has tied similar cases to reverse-engineering efforts to support national defense programs. For example, in March 2013, a representative of a South and Central Asia commercial entity with a history of soliciting cleared contractors for sensitive or classified information and technology emailed a cleared contractor requesting a single pulse modulator.

With so many collection attempts linked to South and Central Asia involving résumé submissions and academic solicitations to cleared contractors engaged in various technological efforts, any specific technology being targeted was often difficult to ascertain. DSS therefore categorized the majority of reported South and Central Asia collection attempts against cleared industry as either unknown or no industrial base technology, at 39 percent and 12 percent, respectively, of FY13 industry submissions.

The listing of identifiable technologies South and Central Asia entities sought from industry in FY13, based on industry reporting, shifted slightly from last fiscal year, perhaps due in part to the change in DSS' categorization methodology. Electronics technology remained the most targeted technology sector, but declined from representing 15 to nine percent of relevant contact reports. Sought-after electronics included space-qualified components and enabling components having a variety of applications, including in radars.

The second most requested technologies, at five percent of the total, fell into the C4 category, formerly part of the larger IS category that in FY12 reporting accounted for 13 percent of the total. Radars, once part of the MCTL's LOS sector, became the third most sought after technology category on their own, tied at four percent of reported contacts

Figure 15: Top Targeted Technologies

 Electronics 9%
Radiation-hardened integrated circuits; enabling components (radar); noise diodes; low noise amplifiers

 Command, Control, Communication, & Computers 5%
Electronic warfare platforms; communication intercept and jamming; global positioning system jamming

 Radars 4%
Fire finder radar; radar test sets

 Nanotechnology 4%
Associated research

with a new category, nanotechnology. Additional cases of identified technologies saw South and Central Asia entities repeatedly expressing interest in classified programs involving aeronautic and marine systems, each at three percent. Reported incidents not involving academic solicitation and seeking employment focused on electronics and C4. Consistent with last year's reporting, South and Central Asia entities generally requested enabling technologies as opposed to whole platforms or systems.

Both in number of reported cases and percentage of the total, the space systems technology sector declined from FY12 to FY13. However, there is a great deal of overlap between space systems and electronics, which remained the most targeted category. Past reporting shows that militaries across the globe are becoming increasingly dependent on space assets—and increasingly concerned by the perceived threat from other states' counterspace systems. Satellite platforms can provide dedicated surveillance, a capability of interest to multiple South and Central Asia states worried about their borders with neighbors. The weight of available evidence indicates that South and Central Asia entities, in an effort to further develop indigenous components while enhancing their countries' space efforts, will seek space electronics developed and produced in the United States, considered the world leader.

Industry reporting also revealed a number of academic solicitations to cleared facilities specializing in space-qualified electronics. For instance, in October 2013, a senior scientist at a cleared facility received separate email requests for summer internships from two South and Central Asia students attending technical institutions.

The scientist's specialty involved survivability of electronics in space, an area in which both students expressed interest in their solicitations.

Analyst Comment: Considering the current general emphasis on both advancing space capabilities and developing indigenous expertise, South and Central Asia entities very likely attempt to obtain information regarding space electronics via any means available, including by attaining positions at cleared facilities. (Confidence Level: Moderate)

A number of South and Central Asia entities sought to enhance relations with cleared contractors dealing in C4 technologies (which accounted for five percent of FY13 industry reporting), specifically EW platforms. This could represent South and Central Asia armies' efforts to upgrade their EW forces due to tensions with rival neighbors. Desired technological fixes included communication intercept and jamming and global positioning system (GPS) jamming. As neighboring rivals' EW systems continue to modernize, South and Central Asia militaries perceive a requirement to upgrade their own capabilities to keep pace.

In FY13, South and Central Asia entities requested a variety of radar systems, the third most targeted technology sector. Relevant radar system functions include command and control, airborne battle management, surveillance of fixed and moving ground targets, enemy situation analysis, and location, tracking, targeting, and attack operations. The systems sought often matched tenders issued by regional armed forces that past reporting indicates are attempting to improve and expand their capabilities by obtaining advanced platforms and weapons to support their perceived defense missions. Some of the radar systems South and Central Asia commercial companies sought were controlled under the International Traffic in Arms Regulations (i.e., ITAR).

Analyst Comment: South and Central Asia armed forces likely seek foreign materials and technology in order to modernize and expand their capabilities, especially with regard to counterinsurgency operations and border monitoring. Collection entities very likely seek the requested radar technologies to upgrade current systems and platforms. (Confidence Level: Moderate)

Marine systems were another target for South and Central Asians soliciting cleared contractors for research positions and postdoctoral degrees. Multiple littoral states admit their desire to dominate the Indian Ocean region and counter any current or emerging threats close to their coastlines. The IC has found that such states are currently engaged in aggressive efforts to modernize their naval capabilities in order to do so. Targeted cleared contractor programs were often involved in underwater acoustics or fluid dynamics; the latter field has aeronautical applications as well.

In FY13, South and Central Asia-connected collectors constituted the fourth most active region at targeting inertial navigation systems (INS), a subset of the positioning, navigation, and time sector of the IBTL. Their collection activity primarily consisted of academic solicitations that attempted to acquire access to microelectromechanical systems-based guidance systems with integrated GPS. See the special focus area of this publication for further discussion of the INS aspect of South and Central Asia collection activity.

Analyst Comment: Based on overall trends in South and Central Asia-related industry reporting, DSS assesses that the region generally lacks adequate indigenous INS production capability, and regional collectors continue to have collection targets for technology in this sector. (Confidence Level: Moderate)

OUTLOOK

South and Central Asia governments are on record as intending to develop their own robust and sustainable defense industries. DSS assesses that related collection entities will almost certainly continue to target a wide range of sensitive and export-controlled technologies. (Confidence Level: High)

To achieve a more sophisticated defense industrial base, South and Central Asia defense organizations will almost certainly seek to enhance their technology exploitation capabilities via academia and/or reverse-engineering. (Confidence Level: High)

DSS assesses that decreased regional defense budgets may mean fewer overall industry reports attributed to South and Central Asia entities in FY14, but more focused collection activities. Regardless

of states' economic woes, their continuing desire to increase their military capabilities, their ongoing internal counterinsurgency operations, and the threats they perceive from regional rivals will likely result in a continuing intent to acquire foreign, particularly U.S., technology. (Confidence Level: Moderate)

Cleared contractor reception of academic solicitations and résumés will very likely continue to surge. South and Central Asia students will very likely attempt to leverage the collaborative nature of academia to extract whatever information they can. (Confidence Level: High)

DSS assesses that South and Central Asia commercial entities acting as procurement agents will very likely continue to be a mainstay of collection efforts to acquire U.S. technology. These entities will almost certainly continue to use AAT and RFI as their preferred MOs, as reflected in industry reporting, since they generally represent the lowest-risk, highest-gain, and seemingly most legitimate means to attempt to acquire U.S. technology. (Confidence Level: High)

South and Central Asia commercial companies that successfully establish relationships with cleared contractors under the solicitation or marketing services MO are likely to use those relationships as a foundation for seeking controlled technology. (Confidence Level: High)

With relations between South and Central Asia and the United States seemingly improving, increased aid and technology-sharing become more likely. South and Central Asia's improved relations with the United States will also likely contribute to enhanced collaborative relationships between the region's commercial entities and cleared contractors. If monitoring detects more intelligence officers within official visiting delegations, it will likely signify a shift toward attempting to leverage access to cleared facilities to gain unauthorized access to U S. information and technology. (Confidence Level: Moderate)

Cleared industry will very likely continue to receive acquisition requests from South and Central Asia procurement firms for limited and single quantities of U.S. technologies, which can facilitate reverse-engineering. (Confidence Level: High)

As South and Central Asia regimes seek to carry out force-wide modernizations and upgrades, DSS assesses that regional collection efforts will likely continue to prioritize enabling technologies falling within the electronics, C4, and radar categories. (Confidence Level: High)

While South and Central Asia individuals' attempts to obtain employment at cleared facilities will probably extend over various U.S. defense platforms and technologies, programs related to nanotechnology and underwater acoustics will likely remain prominent targets. (Confidence Level: Moderate)

Strengthening relationships between South and Central Asia states and third countries of concern will likely perpetuate the risk of transfer of U.S. technology. South and Central Asia regimes are likely to ultimately share with others any classified or sensitive U.S. information and technology they acquire. Since they may not have defense funds available to purchase technology outright, they may feel the need to leverage their existing and improving access to U.S. technology as a quid pro quo for concessions from their other defense procurement sources. (Confidence Level: Moderate)

Internal and foreign relations issues present on the agendas of South and Central Asia states may not have significant, immediate impacts on technology acquisition. However, domestic strife may worsen in the coming months, which would likely cause internal instability. Internal strife coupled with potential escalation of tensions with neighboring rivals would probably have repercussions for the countries' other external relationships, including but not limited to those with the United States. (Confidence Level: Low)

"Frown! You're on Candid Camera!"

This case study details possible South and Central Asia targeting of a cleared contractor employee who was a manager with a U.S. cleared contractor aircraft program.

In April 2013, a South and Central Asia individual identifying himself only as a journalist approached the cleared contractor employee at a U.S. airport. The journalist took a photo of the employee without consent and then proceeded to ask a number of questions regarding the employee's professional and personal life.

The cleared contractor is beginning foreign sales of the aircraft in question, including to South and Central Asia customers. Export versions incorporate export-licensed U.S. sensors as well as unique features, including South and Central Asia-built subsystems tailored to regional specifications.

Regional rivals' equivalent fleets include different, recently upgraded aircraft. However, the IC assesses that acquisition of the targeted aircraft will allow its possessor to surpass its rivals and deploy the most capable aircraft fleet of its type in the region.

Analyst Comment: While regional rivals very likely seek to match each other's advances in aircraft procurement, current defense budget constraints will probably leave some of them without the financial wherewithal to purchase the new aircraft. Therefore, South and Central Asia collectors are likely attempting to acquire information and technology about the cleared contractor's aircraft to contribute to the modernizing and upgrading of their countries' current systems. (Confidence Level: Moderate)

While the alleged journalist's approach may have been a legitimate attempt to acquire information, there is an even chance he targeted the cleared contractor employee due to the employee's association with the aircraft program, hoping to acquire information on the aircraft and/or establish a relationship with the employee. South and Central Asia entities would likely have used any information the journalist garnered from the employee to bolster indigenous production capabilities and develop a better understanding of opponents' aircraft fleet capabilities. (Confidence Level: Moderate)

EUROPE & EURASIA

OVERVIEW

In fiscal year 2013 (FY13), industry reports of foreign collection attempts originating in Europe and Eurasia to obtain unauthorized access to sensitive or classified information and technology increased by nearly 60 percent from FY12, and the region's share of the year's reporting increased as well. Multiple countries within the region continued to be active and frequent attempted collectors. However, Europe and Eurasia remained the fourth most reported region, responsible for 11 percent of total reporting.

Traditionally, a number of the world's most advanced and efficient militaries have resided in Europe and Eurasia, and the region is home to several current campaigns to upgrade and update national militaries to keep this so. These military modernization efforts vary in scale from modest to major. Approaches differ as well, from substantial investments in indigenous military research and development (R&D), to acquisition via outright foreign purchase, to some component of illicit acquisition of technology resident in the U.S. cleared industrial base. Europe and Eurasia's militaries aspire to a range of top-quality marine, aviation, space, and unmanned weapons systems. Almost all of them seek a force structure involving fewer but more professional soldiers wielding high-tech weapons.

Therefore militaries from Europe and Eurasia tend to be interested in cutting-edge technologies. The Defense Security Service (DSS) recently began tracking these technologies more particularly by developing and transitioning to the Industrial Base Technology List (IBTL). Such technology sectors range across new fields and materials: nanotechnology, synthetic biology, quantum systems, genetic, virtual, and like fields, writ large. The science in many of these fields is still in the basic research phase, not yet applied to active programs fulfilling military functions. Thus, it is not surprising that the most notable changes in industry reporting on Europe and Eurasia from FY12 to FY13 were sharp increases in the individual affiliation and the seeking employment method of operation (MO), as collectors sought to "get in on the ground floor" with regard to the development of new military technology.

Beyond that, there was considerable consistency in industry reporting on Europe and Eurasia. Commercial remained the top collector affiliation at 38 percent, although it and all other affiliations lost share to the individual category, which was second at 31 percent, up from 15 percent the previous year. No other affiliation reached 20 percent of the total.

Similarly, while attempted acquisition of technology (AAT) remained the most commonly reported MO in FY13 industry reporting on Europe and Eurasia, at 27 percent of the total, seeking employment became the second most cited at 24 percent, up from three percent the previous year. The request for information (RFI) accounted for 18 percent, and no other MO rose above ten percent of the total.

Entities from Europe and Eurasia continued to show an interest in a diverse range of technologies. The most commonly targeted IBTL category in FY13 industry reporting was electronics, accounting for 12 percent of the total. This marked an increase over FY12's nine percent share; however, reports of attempted collections against several cutting-edge technologies were missing from industry submissions, leaving open the possibility that Europe and Eurasia collectors have already satisfied their acquisition needs for these technologies or are doing so through third parties rather than directly.

A frequently related category, command, control, communication, and computer (C4) systems (last year a part of the Militarily Critical Technologies List's information systems category), was the second most commonly reported technology sector, with seven percent of industry reporting. Aeronautic systems rounded out the top three targeted areas, accounting for six percent of FY13 reports, a drop from 13 percent last year. Together, these three, electronics, C4, and aeronautic systems, accounted for only a quarter of all reporting, yet no other category accounted for more than four percent. In FY13 industry reporting related to Europe and Eurasia, more than a quarter of sought-after technologies remained unknown.

COLLECTOR AFFILIATIONS

Although the commercial affiliation's share of total industry reporting declined from 43 percent in FY12 to 38, its number of cases increased nearly 40 percent. It is not surprising that commercial remained the most commonly attributed collector affiliation in industry reporting linked to Europe and Eurasia. Many countries from the region are allies of the United States and are similar with regard to level of economic development, industrial infrastructure, and innovative ability. Sharing these congruencies, the most straightforward mode of interaction for Europe and Eurasia commercial companies when investigating desirable technology is to approach their counterparts in the U.S. cleared industrial base directly. However, a portion of the interactions consists of nefarious collection activity, with Europe and Eurasia entities attempting to slip illicit requests in among the large volume of legitimate business exchanges.

Analyst Comment: Commercial entities from the region very likely targeted U.S. technology during FY13 not only to meet domestic expectations but

to incorporate the technology into their foreign military sales to improve their competitiveness on the world market. (Confidence Level: High)

As in FY12, many FY13 reports concerned emails from Europe and Eurasia firms seeking to purchase technology or requesting information. Often, requestors made their approaches on behalf of larger commercial entities that work closely with their governments. For example, in July 2013, a representative of a Europe and Eurasia company emailed a cleared contractor requesting electronic circuits often used in space applications. The representative stated that his company was a primary distributor of foreign electronic components for his country's space industry, providing technology to several companies, all of which work closely with the government on the space program.

Many contacts from Europe and Eurasia commercial entities explicitly stated the end user and/or use. However, several others provided little to no information. In December 2012, a Europe and Eurasia commercial representative contacted a cleared contractor requesting a microwave device used in satellite communications. He provided no information regarding an end use or user, simply stating that his company's previous European provider no longer produced the technology. According to the cleared contractor, the device is

Figure 16: Collector Affiliations

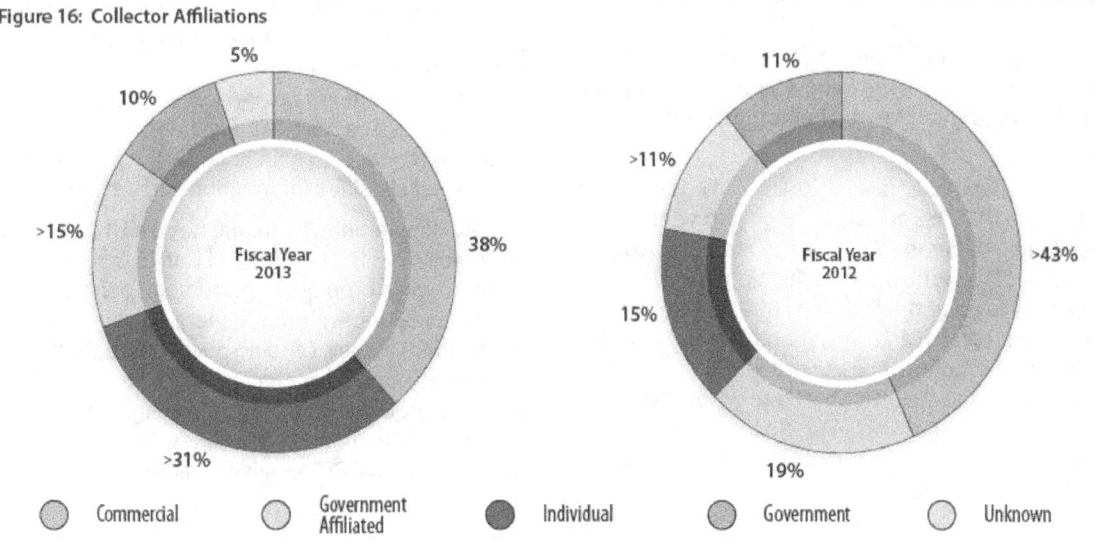

controlled under the International Traffic in Arms Regulations (ITAR) when used for a military-related end.

Analyst Comment: DSS assesses that many of these Europe and Eurasia-linked commercial requests were very likely legitimate business inquiries. However, DSS cannot rule out that some of them, in particular those that sought controlled military technologies yet provided little to no end-use information, represented foreign collector efforts to fulfill government acquisition targets. Commercial entity involvement provides opportunities for collectors to further obfuscate attempts to acquire ITAR-controlled or otherwise restricted technologies. Some regional intelligence services may make use of commercial companies to provide cover for intelligence officers (IOs) so as to leverage their positions to meet national political, defense, and technological objectives. (Confidence Level: Moderate)

The individual affiliation increased notably from FY12 to FY13 in industry reporting, becoming the second most commonly attributed. The number of cases more than tripled and the affiliation's share of Europe and Eurasia-linked reporting went from 15 to 31 percent of the total. Many approaches consisted of individuals from Europe and Eurasia seeking employment with cleared contractors or soliciting academic institutions for research and similar positions.

Analyst Comment: While the vast majority of these applications almost certainly represented individuals following personal opportunities, the overall trend very likely signifies an increased risk of foreign collectors gaining access to cutting-edge and emerging military technologies. (Confidence Level: High)

The number of Europe and Eurasia-linked collection attempts that DSS attributed to government-affiliated entities in FY13 industry reporting rose by over 25 percent, but the affiliation's share of the total fell slightly to 15 percent. These entities included research and educational institutions with government connections. Coupled with a small decrease in the share attributed to government organizations, the overall Europe and Eurasia government signature in industry reports registered a decrease from 30 to 25 percent.

Analyst Comment: The decrease in reports attributed to the two government-connected affiliations in FY13 reporting was small. In most instances, Europe and Eurasia entities willingly disclosed their affiliation with government-connected organizations. However, given that the share of industry reporting for which these affiliations accounted did decrease, DSS cannot rule out that some Europe and Eurasia entities may have more successfully obfuscated end-user information. This may have included intelligence services using companies without their knowledge. (Confidence Level: Low)

METHODS OF OPERATION

Based on industry reporting, Europe and Eurasia entities' most commonly used MO in FY13 remained AAT. The number of such cases increased by a third, although the MO's share diminished from 32 percent of the total in FY12 to 27 percent in FY13. Approaches commonly consisted of sending direct, forthright, and specific emails seeking to purchase technology.

Reports of the seeking employment MO experienced a dramatic rise from FY12 to FY13, paralleling the increase previously discussed for the individual affiliation. For some Europe and Eurasia countries with the most advanced militaries and the most competitive industries, this MO constituted the most frequently employed collection method, based on industry reporting. Figures for academic solicitation remained fairly steady from FY12, and the MO remained the fifth most cited within industry statistics. Within the resultant combined flood of submissions to cleared contractors and related academic institutions and personnel, it was very difficult to discern which applications represented nefarious attempts to gain unauthorized access to sensitive or classified information and technology.

After seeking employment, the next most commonly reported MO in FY13 was the RFI, at 18 percent, representing a drop of ten percentage points. Europe and Eurasia entities generally used email for their RFIs as well, sending questions regarding the cleared contractor's technology. The subject matter of inquiries ranged from pricing information to technical details and capabilities. The number of reported cases of RFI remained stable.

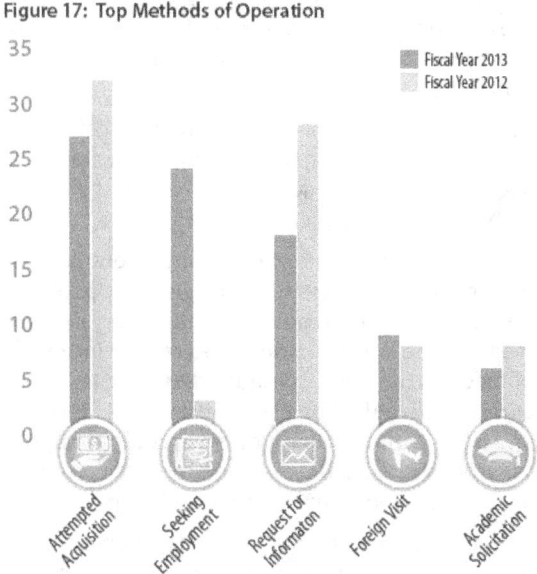

Figure 17: Top Methods of Operation

■ Fiscal Year 2013
■ Fiscal Year 2012

Attempted Acquisition · Seeking Employment · Request for Information · Foreign Visit · Academic Solicitation

Analyst Comment: Most Europe and Eurasia commercial entities very likely view the AAT and RFI MOs as normal, above-board means of attempting to obtain desired information and technology. These requests pose little risk to senders, while even an infrequent success can provide a substantial reward. If direct, overt attempts to purchase a product are effective, there is no need to devote scarce intelligence resources to obtaining the required information or technology. (Confidence Level: Moderate)

The number of industry reports citing the foreign visit MO almost doubled from FY12 to FY13, although the MO's share of total reporting remained stable at nine percent. Visitors to cleared contractors often comprised an official delegation that sometimes included known or suspected IOs. Such IOs are often associated with their country's U.S. embassy and many are effective collectors of economic and technological information, including that resident at cleared contractors.

Analyst Comment: Europe and Eurasia IOs very likely attempt to take advantage of visits to cleared contractors to gather intelligence regarding contractor business activities as well as sensitive or classified information and technology. Some

also very likely attempt to leverage facility visits to learn about contractor employees for possible future exploitation. (Confidence Level: High)

Solicitation or marketing services followed the academic solicitation MO in frequency of industry reporting, at five percent of the total. Some Europe and Eurasia companies contacted cleared contractors with offers to act as their "agent" or "distributor" in the region, while others proposed collaboration on R&D in overlapping areas or joint commercial ventures.

Analyst Comment: Many of these proposals to form business partnerships were probably legitimate. However, such agreements typically involve expectations or even requirements that cleared contractors exchange information and technology with foreign partners and allow delegation visits to cleared facilities, any of which could result in unauthorized access. In addition, foreign collectors likely intend to take advantage of any opportunities such agreements and exchanges confer in order to gain access to sensitive or classified contractor information and technology. (Confidence Level: Moderate)

As in previous years, FY13 industry reporting continued to reflect a relative lack of suspicious network activity linked to Europe and Eurasia: it accounted for only four percent of FY13 industry reporting. Yet the weight of available evidence portrays the region as a repository for advanced abilities in computer-related fields. Within Europe and Eurasia exist communities of skilled cyber operatives, some of whom are hackers embedded in commercial, academic, government, military, and criminal venues. Furthermore, because the region presents a presumption of legitimacy, it offers an attractive location for cyber transfer "hot points" employed to reduce suspicion.

The region maintains highly sophisticated cyber programs and has a long history of conducting cyber operations at a level that presents a persistent and primary threat to U.S. systems. Recent open-source reporting stated that cyber adversaries, including some from Europe and Eurasia, conduct intelligence-collection operations against a variety of global victims. These include, but are not limited to, U.S. and European academic and research institutions as well as U.S. manufacturing and construction industries. During 2013, a cleared contractor constructing an embassy compound for the Department of State reported an attack on its website. The majority of the associated traffic resolved to Europe and Eurasia. However, little is known regarding Europe and Eurasia intent to conduct malware attacks against U.S. systems.

TARGETED TECHNOLOGIES

Entities from Europe and Eurasia targeted electronics; C4; and aeronautic systems in a combined 25 percent of their reported collection attempts. These technologies were the most commonly targeted in overall industry reporting as well.

Electronics was the most targeted technology sector in FY13, accounting for 12 percent of all Europe and Eurasia-related industry reporting. In contrast, in FY12, electronics was only the fourth most reported technology sector, accounting for nine percent. Much of the demonstrated interest in the electronics sector concerned space applications, especially certain types of circuitry in which U.S. cleared contractors remain the world leaders. While Europe and Eurasia's industrial investments and other developmental efforts have allowed some producers to approach the quality and output of their U.S. counterparts, others lag behind, leaving them incapable of meeting internal demands.

Europe and Eurasia entities have accounted for increased numbers of industry-reported attempts to acquire these electronics from abroad, particularly the United States. Industry reporting noted interest in the circuits that was displayed at trade shows and through third countries.

In October 2012, a representative of a commercial company in one Europe and Eurasia country contacted a cleared contractor requesting circuits for ultimate use by a company in a different Europe and Eurasia country.

Reporting from the same month stated that the same representative approached a European employee of an identified U.S. company at a different trade show in Europe. He expressed interest in purchasing certain chips, which were export-

controlled. The representative subsequently called the employee's mobile phone and left messages, and attempted to contact the employee at his house.

Additional reporting from November 2012 noted that different Europe and Eurasia representatives approached the booth of a cleared contractor at another European trade show and expressed interest in the same variety of chips.

Analyst Comment: Europe and Eurasia entities have continued to show interest in the specialized integrated circuits in question, which very likely means some domestic industries remain incapable of meeting the outputs required to achieve current national goals. DSS assesses that Europe and Eurasia entities will almost certainly continue to target specialized U.S. integrated circuits and microprocessors. (Confidence Level: High)

The next most commonly reported technology area was C4, accounting for seven percent of the total. This is frequently an area of attention for countries seeking to upgrade their technology so as to make their armed forces more mobile and adaptable, and to improve their ability to monitor and control military operations. These regimes seek to speed up decision cycles and improve communications security and situational awareness. C4 equipment that lacks security, operates slowly, performs poorly, breaks down frequently, and is difficult to repair impedes achievement of these goals.

Industry reporting regarding Europe and Eurasia C4-related interests showed the targeting of a range of technologies, with some emphasis on secure communications equipment. In June 2013, a cleared contractor received an email from a Europe and Eurasia national requesting two types of encrypted smart phones for secure voice communication. The individual provided inadequate end-use information. The same cleared contractor also received a request in April from a different individual for one unit of the same phone, with no end-use information. Access to such export-controlled communications security (i.e., COMSEC) technology could enable adversaries to perform cryptanalysis against sensitive U.S. cryptographic systems.

Analyst Comment: Although these requests were for a limited number of units, DSS cannot rule out that they represented efforts to enable specific

Figure 18: Top Targeted Technologies

Electronics 12%
Radiation-hardened
integrated circuits;
microelectronics;
microprocessors

**Command, Control,
Communication, &
Computers 7%**
Secure communications
equipment; encrypted
smartphones

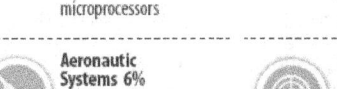

**Aeronautic
Systems 6%**
Unmanned aerial
vehicles; fighter aircraft

Radars 4%
Battlefield radar;
synthetic aperture radar

encrypted communication or cryptanalysis capabilities. Alternatively, the small numbers of units requested may signify an intention to reverse-engineer any technology acquired to benefit Europe and Eurasia cryptographic development or to counter similar U.S. systems. (Confidence Level: Low)

A February 2013 RFI linked to Europe and Eurasia was noteworthy because it dealt with the development of a secure data communications link for use on unmanned aerial vehicles (UAVs). A company representative stated that his Europe and Eurasia company was designing a satellite data link with an antenna unit. It would send secure data and stream video from airborne platforms to ground stations, ostensibly for aerial monitoring of energy-related systems. The representative sought information on similar systems.

Analyst Comment: This request was particularly notable in that it combined interest in improving secure communications with expanding UAV capability. While current Europe and Eurasia commercial endeavors justify expanding capabilities in both areas, DSS cannot discount that any information satisfying this request would have ultimately furthered foreign military goals. (Confidence Level: Moderate)

The aeronautic systems technology sector, referred to above, constituted the third most common technology area Europe and Eurasia collector entities targeted in FY13 industry reporting. In FY13, the sector experienced a decrease in the number of reported cases, and in share of reporting from 13 percent in FY12 to six percent. However, within the aeronautic systems sector, Europe and Eurasia

entities continued to show a notable interest in UAVs. Targets included medium-range models, both small and larger, with internal/external payload capability.

Analyst Comment: There is an even chance that reports of Europe and Eurasia collection efforts against other aeronautics subsectors, such as fixed-wing and rotary aircraft, declined because some of the region's producers feel they have achieved their goals in those areas. If so, DSS assesses that this success very likely resulted from a combination of indigenous development and past successes in obtaining unauthorized access to sensitive or classified information and technology resident in the U.S. cleared industrial base. (Confidence Level: Moderate)

In addition, regarding the aeronautics sector overall, Intelligence Community analysis found that Europe and Eurasia capabilities have benefited from a combination of enhanced commercial involvement and increased international cooperation. Government encouragement of the satisfaction of military prescriptions via international commercial arrangements can tend to retard indigenous development in the short term but contribute to it in the middle and longer term.

Analyst Comment: Governmental prioritization toward technology areas such as UAVs very likely contributed to the expansion of Europe and Eurasia companies operating in that field and, by extension, continued collector interest in obtaining information regarding U.S. systems. DSS assesses that demonstrations of Europe and Eurasia collector entities' interest in UAVs and associated enabling technologies will very likely continue. (Confidence Level: High)

Europe and Eurasia entities in FY13 were not among the top regions reported as targeting U.S. inertial navigation systems (INS) technology. INS fall within the positioning, navigation, and time technology sector of the IBTL, which accounted for only two percent of Europe and Eurasia-connected reports. For more discussion of INS, see the special focus area of this publication.

OUTLOOK

Given the emphasis that Europe and Eurasia regimes are placing on military modernization and restructuring and some demonstrated willingness to integrate foreign technology into their military systems, DSS assesses that Europe and Eurasia collectors will almost certainly continue to target U.S. cleared industry. (Confidence Level: High)

To the extent that Europe and Eurasia industrial capabilities lag behind those of the United States technologically, the region's national leaders almost certainly see obtaining sensitive or classified U.S. information and technology, whether licitly or illicitly, as the fastest way to make up ground. (Confidence Level: High)

DSS assesses that Europe and Eurasia collection efforts will very likely continue to rely on commercial entities, especially since this approach offers at least an appearance of legitimacy and cooperation. Governments will likely continue to encourage and emphasize the role of joint commercial ventures. Such joint ventures as well as the use of commercial front companies and even straightforward commercial approaches all very likely offer opportunities for foreign collectors to attempt to exploit cleared contractors. The volume of industry reporting of collection activity traced to Europe and Eurasia commercial entities will very likely continue to increase. (Confidence Level: High)

AAT and RFI, the first and third most cited MOs in Europe and Eurasia-connected industry reporting in FY13, will very likely continue as preferred MOs, in part because they permit the intermixing of illicit collection attempts with normal commercial interactions. (Confidence Level: Moderate)

As cutting-edge technologies continue to evolve and move toward applications, Europe and Eurasia will very likely continue to be a prominent source of submissions from individuals seeking employment and soliciting academic placement with companies and institutions conducting research and developing applications, both on those individuals' own behalf and for the intended benefit of foreign companies and governments. (Confidence Level: Moderate)

Europe and Eurasia collection efforts will almost certainly continue to be diverse and widespread. However, given the significant recent increase in requests for electronics technology, this sector very likely represents a particularly pressing need, and cleared industry will very likely continue to receive requests for electronics. (Confidence Level: High)

Global economic conditions and the world geopolitical situation have combined to lead to cuts in defense spending both in the United States and Europe. This has had effects on U.S. cleared industry, and will almost certainly have further impacts. DSS assesses that those Europe and Eurasia countries that are still determined to increase their spending on military modernization will probably seek to exploit this situation to increase potential access to sensitive or classified information and technology and thereby strengthen indigenous production capabilities. Collection efforts will probably involve both attempting to purchase technologies outright and encouraging U.S. firms to establish facilities in Europe and Eurasia. (Confidence Level: High)

Other Europe and Eurasia countries represent a risk of diversion of any technology obtained to third countries. Those with weak export-control regimes place resident technology at risk of illicit transfer. To sustain the viability of their defense industries, such countries often make technology transfer a part of foreign military sales. Therefore, transfers of U.S. technology to these countries pose a risk of further proliferation. Cleared industry will likely continue to experience a growing number of suspicious contacts from entities in such countries, primarily commercial. Any technology shared with entities from these countries or obtained via unauthorized access will very likely be transferred to third countries, either illicitly or as part of foreign arms sales. (Confidence Level: High)

To some extent, Europe and Eurasia is experiencing degradation in its security situation, with ongoing challenges and additional possible ones on the horizon. Responding to these challenges as they arise and become acute may result in the eruption of new and intense technology collection campaigns by affected Europe and Eurasia regimes. (Confidence Level: Low)

Getting By With a Little Help from a Friend

The following case study demonstrates how some Europe and Eurasia collection efforts use small companies to attempt to obtain technology on behalf of larger companies that have connections to a government. In this instance, the request was for sensitive electronics equipment that has various space and military applications.

In January 2013, a representative of a Europe and Eurasia company requested from a cleared contractor an export-controlled amplifier on behalf of another company from the same country. At the first company's request, the second company later provided an end-user certificate identifying the eventual recipient as a third entity that directly supplies navigational equipment to naval vessels.

The same representative contacted the same cleared contractor again in March 2013, requesting various export-controlled electronics components, also on behalf of the second company. According to the end-use certificate that company provided, the components were to be integrated into test, telecommunications, and perhaps electronic warfare (EW) equipment.

Only limited information was available to substantiate the identity of the various purported Europe and Eurasia end users. No evidence corroborated the existence of the first company. Further, although its nominal representative provided the second company's website link, that webpage provided only minimal details regarding the company's business scope and contact information. The website claimed the company develops circuitry and cabling equipment; notably, the site pictures a warship.

Analyst Comment: These requests provided information demonstrating a Europe and Eurasia interest in communication and EW systems, as well as the use of commercial entities for attempted acquisition. In addition, both instances demonstrated at least some effort to obfuscate the ultimate end user. The third entity's provision of integration services to its country's navy suggests the end users would have put the requested technology to military use, likely for direction-finding or EW. (Confidence Level: Moderate)

OTHER REGIONS

In fiscal year 2013 (FY13), entities associated with the Western Hemisphere and Africa together accounted for seven percent of the suspicious activities industry reported to the Defense Security Service (DSS). This reversed a three-year trend that saw these two regions accounting for a diminishing share of industry reporting. In FY09, DSS identified entities from these regions in ten percent of industry reporting. From FY09 through FY12, this share of industry reporting declined to just under five percent.

WESTERN HEMISPHERE

Of the two regions, Western Hemisphere entities accounted for most of the FY13 increase in associated reporting. They accounted for six percent of the suspicious incidents industry reported to DSS, up from just four percent in FY12. The number of incidents DSS attributed to Western Hemisphere entities more than doubled in FY13 compared to FY12. Even with this marked increase, reported activity associated with entities from this region totaled only 51 percent of that of the fourth most active region, Europe and Eurasia.

Analyst Comment: Collectors from other regions often use front companies or brokers operating in the Western Hemisphere to target U.S. technologies. It is likely that many of the reported incidents attributed to entities in the Western Hemisphere actually constituted the front end of foreign end users' collection attempts. (Confidence Level: High)

Since FY08, commercial has been the most common affiliation of collectors originating in the Western Hemisphere, accounting for 48 percent of relevant data in FY13. Based on industry reporting, commercial entities relied heavily on three methods of operation (MO): request for information, attempted acquisition of technology, and solicitation or marketing services. While these three MOs accounted for 66 percent of suspicious activity attributed to all entities from this region, commercial entities used these MOs in 94 percent of their attempts to obtain unauthorized access to sensitive or classified information and technology resident in the U.S. cleared industrial base.

Based on industry reporting, collectors from the Western Hemisphere primarily targeted the same three technology sectors as those from the four most prolific collector regions: electronics; aeronautic systems; and command, control, communication, and computers (C4). In FY13, electronics was the targeted technology in 17 percent of reported suspicious incidents attributed to entities from this region, while aeronautic systems and C4 were each targeted in nine percent. This is the second year electronics was the most targeted technology sector.

Figure 19: Top Targeted Technologies

Electronics 17%
Radiation-hardened integrated circuits; KA-band gallium nitride amplifiers; monolithic microwave integrated circuits, traveling wave tubes

Aeronautic Systems 9%
Unmanned aerial systems; helicopter components; display panels (up front control display & heads up display)

Command, Control, Communication, & Computers 9%
Iridium satellite phone; tactical/handheld/mobile radios; antennas (spiral antennas); cryptographic modules

Manufacturing Equipment & Processes 7%
Ferrous & non-ferrous mechanical components; semiconductor packaging; circuit board assembly solutions

Figure 20: Top Methods of Operation

Fiscal Year 2013
Fiscal Year 2012

Request for Information; Solicitation or Marketing Services; Attempted Acquisition; Seeking Employment; Academic Solicitation

AFRICA

Since at least FY07, collectors from Africa have been the least active among the regions. During this period, they accounted for no more than two percent of the suspicious activity cleared industry reported. In FY13, DSS attributed less than one percent of reported suspicious activities to entities from Africa. However, the number of collection attempts attributed to this region continues to increase each year. The number of reported collection attempts by entities from this region increased 57 percent in FY13 over FY12 and by 94 percent in FY12 over FY11.

In FY13, individual was the most common affiliation for collectors from Africa. DSS attributed 40 percent of reported suspicious contacts associated with this region to individuals. We attributed a further 36 percent of industry reports to commercial entities. From FY10 through FY12, commercial had been the most common affiliation, with individuals accounting for no more than 18 percent of reported suspicious incidents. The most common MO for entities from Africa in FY13 data was seeking employment, cited in just over half of reported suspicious contacts. In contrast, from FY09 through FY12, this was the attributed MO in less than four percent of reported suspicious contacts.

Entities linked to Africa most commonly targeted aeronautic systems, electronics, C4, software, optics, space systems, and directed energy technologies. Combined, all seven of these technology sectors accounted for just 45 percent of the Africa-linked suspicious incidents industry reported. Based on

industry reporting, since FY09 aeronautic systems has been the technology sector collectors from this region targeted first or second most.

In FY13, the targeted technology was unknown in 47 percent of the incidents attributed to collectors from Africa. The large number of incidents for which the targeted technology was unknown corresponded with the advent of individual as the most common collector affiliation and seeking employment as the most common MO. In 89 percent of reported cases of seeking employment, the targeted technology was unknown, since applicants seeking employment in cleared industry often did not specify any U.S. technology programs in their résumés or curricula vitae.

Analyst Comment: The number of reported targeting attempts against U.S. technology originating from or directed through the Western Hemisphere and Africa will probably continue to increase. However, entities from these two regions will very likely remain less active than entities from the other four regions, probably accounting for no more than ten percent of reported suspicious activity combined. (Confidence Level: Moderate)

Figure 21: Top Targeted Technologies

Aeronautic Systems 16%
Scan Eagle UAV; RQ11B Raven UAV; Hybrid Quadrotor UAV; Bell 212 helicopters

Electronics 15%
Dimmable infrared LED indicators; Bandit dual band transceivers; vehicle scanning system

Command, Control, Communication, & Computers 4%
Fiber telecommunications; KA-band satellite communications; Stingray II

Software 4%
T-Rex software; SpecTRM software; modeling & simulation software

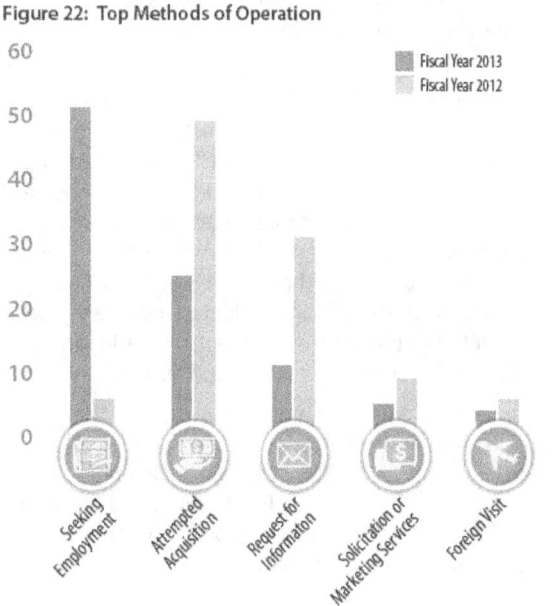

Figure 22: Top Methods of Operation

Fiscal Year 2013
Fiscal Year 2012

Seeking Employment | Attempted Acquisition | Request for Information | Solicitation or Marketing Services | Foreign Visit

CONCLUSION & OUTLOOK

The tide of foreign targeting of U.S. technologies shows no sign of ebbing. The Defense Security Service (DSS) received and reviewed over 30,000 reports from cleared industry in fiscal year 2013 (FY13). DSS referred 5,448 of these reports to other government agencies. Those agencies initiated 717 investigations or operations based on DSS referrals. The number of reports from cleared industry that DSS considered to be of counterintelligence (CI) concern, and to likely represent an individual's efforts to obtain unauthorized access to sensitive or classified information and technology, increased by 33 percent in FY13 over the previous year.

The geographic regions from which these collection attempts originated retained the same hierarchy, based on frequency of citation in industry reports, in FY13 as in FY12. Since FY07, East Asia and the Pacific has been the most active collector region, the Near East the second most active. In FY13, these two regions accounted for 62 percent of all suspicious incidents industry reported that DSS considered to be of CI concern. Since FY07, these two regions collectively have accounted for at least 56 percent of annual collection activity.

The meteoric growth from FY10 through FY12 in collection attempts attributed to entities from East Asia and the Pacific slowed in FY13. In FY13, the number of collection attempts DSS attributed to entities from this region increased by just 15 percent over FY12. FY09 was the last year during which reported activity linked to East Asia and the Pacific grew at a slower rate than in FY13. Still, in FY13 entities from the region conducted more reported collection attempts than entities originating in the second and third most active regions combined.

Government was the most common affiliation for collection entities from East Asia and the Pacific, credited with 28 percent of reported activity, down from 41 percent in FY12. Suspicious network activity (SNA) remained the most common method of operation (MO) for entities from East Asia and

the Pacific. Government entities conducted nearly 79 percent of the reported SNA originating in this region.

Entities from East Asia and the Pacific targeted a wide array of technology in FY13, encompassing 27 of the 29 technology categories in the Industrial Base Technology List (IBTL). Entities from this region most frequently targeted electronics; command, control, communication, and computers (C4); aeronautic systems; marine systems; and software. These targeting priorities have remained consistent over the past seven years. For the second year running, electronics was the most targeted technology sector, and it has been one of the five most targeted sectors for entities from this region since FY08.

This is the first year DSS used the IBTL. In previous years, C4 and software together constituted the information systems (IS) category. In FY07 through FY11 industry reporting, IS was the technology sector that East Asia and the Pacific entities targeted most frequently, and was the second most targeted in FY12. If still combined, C4 and software would have been the most targeted technology category in FY13. Aeronautic systems, the third most targeted technology in FY13 by collectors from this region, based on industry reporting, has been among the top five most targeted technologies since FY07.

Collection attempts DSS attributed to entities from the Near East increased by 52 percent in FY13 over FY12. For the third year, government-affiliated was the most common affiliation for Near East entities industry reported as attempting to collect information and technology from cleared facilities. Commercial remained the second most commonly reported collector affiliation for Near East entities, with those entities accounting for 22 percent of reports related to the region. This was the smallest share of activity attributed to commercial entities from any of the six regions. Academic solicitation was the MO Near East entities used most, based on industry reporting. Near East entities continued to seek a wide variety of military and dual-use technologies. The most commonly reported were marine systems technology, electronics, and aeronautic, space, and energy systems.

Analyst Comment: Near East collectors' extensive use of procurement networks likely concealed some of their activity, contributing to the commercial affiliation continuing to account for a relatively low proportion of reporting attributed to the region. Commercial companies from the region that, for a variety of reasons, could not make direct approaches to cleared contractors very likely originated many of their requests from other countries having more favorable trade relationships with the United States. (Confidence Level: High)

For the second year running, South and Central Asia and Europe and Eurasia were the third and fourth most active collector regions, based on industry reporting. In FY10 and FY11, these positions were reversed. In addition, for the seventh consecutive year, the Western Hemisphere and Africa were the least frequently identified regions of origin for collection attempts targeting information and technology resident in cleared industry.

Analyst Comment: In the coming years, the collector regions will almost certainly remain consistent in relative levels of reported activity, although some minor shifting of rankings may occur. East Asia and the Pacific will almost certainly remain the most active collector region, the Near East the second most active. South and Central Asia will very likely remain the third most prolific collector region, followed by Europe and Eurasia. (Confidence Level: High)

For the seventh consecutive year, commercial was the most reported affiliation of entities targeting U.S. technologies. However, for the fourth year, its share of reported collection attempts diminished. In FY09, commercial entities accounted for 49 percent of all suspicious contacts cleared industry reported, whereas in FY13 DSS attributed 27 percent to commercial entities, only slightly greater than

the share it attributed to government-affiliated entities. Since FY09, DSS has attributed an increasing percentage of reported suspicious incidents to government-affiliated entities.

Academic solicitation was the preferred MO for collectors in FY13 data. Collectors used it in 22 percent of reported collection attempts. Attribution of the academic solicitation MO increased by over 81 percent in FY13. It was the most commonly reported MO for entities from the Near East, and the second most for entities from East Asia and the Pacific and South and Central Asia.

Foreign collectors used SNA in 19 percent of collection attempts targeting cleared industry, based on FY13 industry reporting. By contrast, SNA was the preferred MO in FY12, identified in 29 percent of reported collection attempts. Collectors from East Asia and the Pacific remained the preeminent users of SNA. They accounted for over 67 percent of SNA in overall FY13 industry reporting and 72 percent in FY12. Due to the nebulous nature of the cyber threat, DSS could not identify a region of origin in nearly 25 percent of reported SNA cases.

Analyst Comment: The number of reports citing evidence of SNA will likely return in FY14 to its previous higher level. However, DSS cannot dismiss the possibility that the number will continue to fall. Cleared contractors have improved their ability to detect and defeat cyber attacks, assisted by both government information and reports from private firms. (Confidence Level: High)

Notwithstanding these defensive efforts, cyber actors will very likely continue to conduct spear-phishing attacks and attempt network intrusions, and continue to both adjust existing exploitation techniques and develop new ones. Thus, SNA will very likely remain a significant threat to cleared industry networks. Entities originating in East Asia and the Pacific will almost certainly remain the preeminent users of SNA in targeting cleared industry. (Confidence Level: High)

> DSS received and reviewed over 30,000 reports from cleared industry in FY13. DSS referred 5,448 of these reports to other government agencies. Those agencies initiated 717 investigations or operations based on DSS referrals.

Attempted acquisition of technology (AAT) remained one of the top three MOs for targeting technology resident in cleared industry, based on industry reporting. DSS identified AAT as the MO in 15 percent of reported FY13 collection attempts. This continued a downward trend for AAT: it was the preferred MO in FY11 data, used in 23 percent of reported suspicious contacts, and the second most used in FY12, identified in 20 percent of reports that year.

Analyst Comment: Although the exact rates at which foreign entities employ different MOs may vary, those MOs most frequently reported in the recent past will likely remain so. Academic solicitation, SNA, AAT, and request for information have topped the list for the last three years. These MOs offer minimal risk and low cost, and foreign entities targeting cleared industry will almost certainly continue to employ them when attempting to obtain unauthorized access to sensitive or classified U.S. information and technology. (Confidence Level: High)

DSS' transition from using the Militarily Critical Technologies List (MCTL) to the IBTL created multiple new technology categories from what had been single MCTL categories. IS was the most targeted MCTL category in FY12, but, as noted previously, the IBTL divides IS into the separate categories of C4 and software. Collectively, C4 and software remained the most targeted technology in FY13 reports of collection attempts, identified in nearly nine percent.

Electronics was the most sought-after technology for foreign collectors targeting cleared industry in FY13, based on industry reporting. However, the number of reports of foreign targeting of electronics technology actually dropped by 11 percent compared to FY12. The next most commonly targeted technologies in FY13 were C4, aeronautic systems, marine systems, and software, with each accounting for at least three percent of reported collection attempts.

Lasers, optics, and sensors (LOS) was one of the three most targeted MCTL technologies from FY10 through FY12. For FY13, the IBTL separated LOS into four categories: radars, optics, sensors (acoustic), and lasers. Collectively, these four technology categories accounted for seven percent of FY13 reports of suspicious incidents targeting U.S. technologies, whereas in FY12 LOS accounted for ten percent.

Analyst Comment: Entities from multiple regions target U.S. technologies to support military modernization programs. These modernization programs will likely drive the targeting of marine, space, C4, and electronics technologies to fill perceived military technology gaps. Indigenous industries in these regions will likely use any technologies obtained to further their research and development and reverse-engineering programs, and transfer technology to third countries as part of arms sales. (Confidence Level: Moderate)

East Asia and the Pacific will very likely remain the most reported collector region for targeting technologies involved with inertial navigation systems (INS). Collectors from this region will likely rely heavily on AAT to attempt to obtain access to INS component technology. (Confidence Level: High)

Industry reported few suspicious contacts relating to the emerging technologies of computational modeling of human behavior, quantum systems, synthetic biology, or cognitive neuroscience. Collectively they accounted for less than two-tenths of one percent of overall industry reporting in 2013. Among these technologies, computational modeling of human behavior was the most targeted. Collection activity originating in East Asia and the Pacific accounted for 50 percent of the reported incidents targeting these emerging technologies, while collectors from South and Central Asia and the Near East accounted for 30 and 20 percent, respectively.

Analyst Comment: Only a limited number of foreign governments and private enterprises have mature programs dealing with these technologies. The science that undergirds these technology categories largely remains in the basic, theoretical research phase, so is usually not yet tied to a classified contract at a cleared facility. This likely contributed to the low rate of reported targeting. (Confidence Level: Moderate)

As cleared industry begins to apply emerging technologies to military and commercial programs, the top collectors against U.S. information and technology will almost certainly increase their efforts targeting those sectors. Countries targeting these technologies will almost certainly employ various MOs as they attempt to keep pace with the rapid advancement of U.S. technology for military application. Among technologies DSS categorizes as emerging, quantum computing and quantum key encryption will likely be the most sought-after. (Confidence Level: High)

DSS analysis of FY13 cleared industry reporting demonstrated that foreign entities continue to increase their targeting of U.S. technologies. Foreign entities show all signs of continuing to target U.S. technologies just as vigorously, if not more so, in FY14. Maintaining the U.S. military, technological, and economic edge rests on cleared contractors practicing continued vigilance to ensure that information and technology resident in their facilities is safeguarded.

ACRONYMS & ABBREVIATIONS

A2/AD	anti-access and area denial
AAT	attempted acquisition of technology
ANV	assessed no value
C4	command, control, communication, and computers
CFD	computational fluid dynamics
CI	counterintelligence
CNE	computer network exploitation
CNO	computer network operations
COMSEC	communications security
DoD	Department of Defense
DSS	Defense Security Service
DTG	dynamically tuned gyroscope
ERC	End-User Review Committee
EW	electronic warfare
FOG	fiber optic gyroscope
FY	fiscal year
G&C	guidance and control
GPS	global positioning system
IBTL	Industrial Base Technology List
IC	Intelligence Community
IMU	inertial measurement unit
INS	inertial navigation system
IO	intelligence officer
IS	information systems
IT	information technology
ITAR	International Traffic in Arms Regulations

LOS	lasers, optics, and sensors
M&S	modeling and simulation
MCTL	Militarily Critical Technologies List
MEMS	microelectromechanical system
MMIC	monolithic microwave integrated circuit
MO	method of operation
NISPOM	National Industrial Security Program Operating Manual
PDF	portable document format
PNT	positioning, navigation, and time
QFA	quartz flexure accelerometer
R&D	research and development
rad-hard	radiation-hardened
RFI	request for information
RI	research institute
RLG	ring laser gyroscope
SCR	suspicious contact report
SIGINT	signals intelligence
SME	subject-matter expert
SNA	suspicious network activity
SOE	state-owned enterprise
TTPs	tactics, techniques, and procedures
UAS	unmanned aerial system
UAV	unmanned aerial vehicle
UCR	unsubstantiated contact report
WMD	weapons of mass destruction

INDUSTRIAL BASE TECHNOLOGY LIST

The Defense Security Service's (DSS) Industrial Base Technology List (IBTL) added six categories not reflected in the Militarily Critical Technologies List (MCTL) for emerging technologies, as defined by the Department of Defense, that reside in facilities over which DSS has security oversight responsibilities under the National Industrial Security Program. For example, the IBTL adds agriculture technology because DSS is the cognizant security agency for Department of Agriculture cleared facilities. The IBTL also captures emerging technologies being researched in cleared industry.

MILITARILY CRITICAL TECHNOLOGIES LIST

Information Systems

Information Security

Lasers, Optics, & Sensors

Ground Systems

Aeronautic Systems

Marine Systems

Space Systems

Nuclear Systems

Chemical

Biological

Biomedical

Armaments & Energetic Materials

Directed Energy Systems

INDUSTRIAL BASE TECHNOLOGY LIST

Command, Control, Communication, & Computers

Software

Lasers

Optics

Sensors (Acoustic)

Radars

Ground Systems

Aeronautic Systems

Marine Systems

Space Systems

Nuclear

Chemical

Biological

Medical

Armament & Survivability

Energetic Materials

Directed Energy

	Energy Systems
	Electronics
Energy Systems	Manufacturing Equipment & Processes
Electronics	
Processing & Manufacturing	Positioning, Navigation, & Time
Positioning, Navigation, & Time	
Signature Control	Signature Control
Materials & Processes	Materials (Raw & Processed)
	Synthetic Biology
New Categories	Nanotechnology
	Agricultural
	Cognitive Neuroscience
Weapons Effects*	Computational Modeling of Human Behavior
	Quantum Systems

* The weapons effects category is now part of the nuclear, chemical, and biological categories.

REGIONAL
BREAKDOWN

Africa	East Asia & the Pacific	Europe & Eurasia	Near East	South & Central Asia	Western Hemisphere
Angola	Australia	Albania	Algeria	Afghanistan	Antigua and Barbuda
Benin	Brunei	Andorra	Bahrain	Bangladesh	Argentina
Botswana	Burma	Armenia	Egypt	Bhutan	Aruba
Burkina Faso	Cambodia	Austria	Iran	India	Bahamas, The
Burundi	China	Azerbaijan	Iraq	Kazakhstan	Barbados
Cameroon	Fiji	Belarus	Israel	Kyrgyzstan	Belize
Cabo Verde	Indonesia	Belgium	Jordan	Maldives	Bermuda
Central African Republic	Japan	Bosnia and Herzegovina	Kuwait	Nepal	Bolivia
Chad	Kiribati	Bulgaria	Lebanon	Pakistan	Brazil
Comoros	Korea, North	Croatia	Libya	Sri Lanka	Canada
Congo, Democratic Republic of the	Korea, South	Cyprus	Morocco	Tajikistan	Cayman Islands
Congo, Republic of the	Laos	Czech Republic	Oman	Turkmenistan	Chile
Cote d'Ivoire	Malaysia	Denmark	Palestinian Territories	Uzbekistan	Colombia
Djibouti	Marshall Islands	Estonia	Qatar		Costa Rica
Equatorial Guinea	Micronesia, Federated States of	Finland	Saudi Arabia		Cuba
Eritrea	Mongolia	France	Syria		Curacao
Ethiopia	Nauru	Georgia	Tunisia		Dominica
Gabon	New Zealand	Germany	United Arab Emirates		Dominican Republic
Gambia, The	Palau	Greece	Yemen		Ecuador
Ghana	Papua New Guinea	Holy See			El Salvador
Guinea	Philippines	Hungary			Grenada
Guinea-Bissau	Samoa	Iceland			Guatemala
Kenya	Singapore	Ireland			Guyana
Lesotho	Solomon Islands	Italy			Haiti
Liberia	Taiwan	Kosovo			Honduras
Madagascar	Thailand	Latvia			Jamaica
Malawi	Timor-Leste	Liechtenstein			Mexico
Mali	Tonga	Lithuania			Nicaragua
Mauritania	Tuvalu	Luxembourg			Panama
Mauritius	Vanuatu	Macedonia			Paraguay
Mozambique	Vietnam	Malta			Peru
Namibia		Moldova			St. Kitts and Nevis
Niger		Monaco			St. Lucia
Nigeria		Montenegro			St. Maarten
Rwanda		Netherlands			St. Vincent and the Grenadines
Sao Tome and Principe		Norway			Suriname
Senegal		Poland			Trinidad and Tobago
Seychelles		Portugal			United States
Sierra Leone		Romania			Uruguay
Somalia		Russia			Venezuela
South Africa		San Marino			
South Sudan		Serbia			
Sudan		Slovakia			
Swaziland		Slovenia			
Tanzania		Spain			
Togo		Sweden			
Uganda		Switzerland			
Zambia		Turkey			
Zimbabwe		Ukraine			
		United Kingdom			

REFERENCES

1. Journal article; Journal of Theoretical and Applied Mechanics; Strapdown Inertial Navigation Systems; Albert Ortyl and Zdzislaw Gosiewski; Jan 1998; pages 81–95

2. Journal article; Navigation: Journal of the Institute of Navigation; History of Inertial Navigation; W. Wrigley; Vol. 24, No. 1; Spring 1977; pages 1–6

3. Book; Strapdown Inertial Navigation Technology, 2nd edition; David Titterton and John Weston, 1996